ORGANIZATIONAL ACCIDENTS

To Thomas Augustus Reason (1879–1958).
My grandfather to whom I owe much more than my existence.

Organizational Accidents Revisited

JAMES REASON
Professor Emeritus, University of Manchester, UK

ASHGATE

Published by
Ashgate Publishing Limited
Wey Court East
Union Road
Farnham
Surrey, GU9 7PT
England

Ashgate Publishing Company
110 Cherry Street
Suite 3-1
Burlington, VT 05401-3818
USA

www.ashgate.com

British Library Cataloguing in Publication Data
A catalogue record for this book is available from the British Library

The Library of Congress has cataloged the printed edition as follows:
A catalogue record for this book is available from the Library of Congress

ISBN 9781472447654 (hbk)
ISBN 9781472447685 (pbk)
ISBN 9781472447661 (ebk – PDF)
ISBN 9781472447678 (ebk – ePUB)

Printed in the United Kingdom by Henry Ling Limited, at the Dorset Press, Dorchester, DT1 1HD

Contents

List of Figures and Tables

Figures

Tables

About the Author

James Reason is Professor Emeritus of Psychology at the University of Manchester, England. He is consultant to numerous organizations throughout the world, sought after as a keynote speaker at international conferences and author of several renowned books including *Human Error* (CUP, 1990), *Managing the Risks of Organizational Accidents* (Ashgate, 1997), *The Human Contribution* (Ashgate, 2008) and *A Life in Error* (Ashgate, 2013).

Chapter 1
Introduction

The term 'organizational accidents' (shortened here to 'orgax') was coined in the early 1990s and was developed in the Ashgate book published in 1997 entitled *Managing the Risks of Organizational Accidents*. Sales of the book indicate that many people may have read it – it remains Ashgate's all-time best-selling book on human performance. So the present book is not a revision, but a revisit. A lot has happened in the ensuing 18 years, my aim here is to update and extend the arguments presented in the first book to accommodate these developments. In short, this book is an addition rather than a replacement. And enough has happened in the interim to require a separate book.

Despite their huge diversity, each organizational accident has at least three common features: hazards, failed defences and losses (damage to people, assets and the environment). Of these, the most promising for effective prevention are the failed defences. Defences, barriers, safeguards and controls exist at many levels of the system and take a large variety of forms. But each defence serves one or more of the following functions:

- to create understanding and awareness of the local hazards;
- to give guidance on how to operate safely;
- to provide alarms and warnings when danger is imminent;
- to interpose barriers between the hazards and the potential losses;
- to restore the system to a safe state after an event;
- to contain and eliminate the hazards should they escape the barriers and controls;
- to provide the means of escape and rescue should the defences fail catastrophically.

These 'defences-in-depth' make complex technological systems, such as nuclear power plants and transport systems, largely proof against single failures, either human or technical. But no defence is perfect. Each one contains weaknesses, flaws and gaps, or is liable to absences. Bad events happen when these holes or weaknesses 'line up' to permit a trajectory of accident opportunity to bring hazards into damaging contact with people and/or assets. This concatenation of failures is represented diagrammatically by the Swiss cheese model (Figure 1.1) – to be reconsidered later.

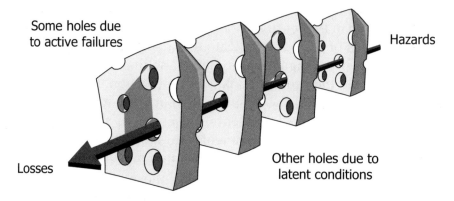

Successive layers of defences, barriers, and safeguards

Figure 1.1   The 'Swiss cheese' model of accident causation

The gaps in the defences arise for two reasons – *active failures* and *latent conditions* – occurring either singly or in diabolical combinations. They are devilish because in some cases the trajectory of accident liability need only exist for a very short time, sometimes only a few seconds:

> *Active failures*: these are unsafe acts – errors and/or procedural violations – on the part of those in direct contact with the system ('sharp-enders'). They can create weaknesses in or among the protective layers.
>
> *Latent conditions*: in earlier versions of the Swiss cheese model (SCM), these gaps were attributed to latent failures. But there need be no failure involved, though there often is. A condition

is not necessarily a cause, but something whose presence is necessary for a cause to have an effect – like oxygen is a necessary condition for fire, though an ignition source is the direct cause.

Designers, builders, maintainers and managers unwittingly seed latent conditions into the system. These arise because it is impossible to foresee all possible event scenarios. Latent conditions act like resident pathogens that combine with local triggers to open up an event trajectory through the defences so that hazards come into harmful contact with people, assets or the environment. In order for this to happen, there needs to be a lining-up of the gaps and weaknesses creating a clear path through the defences. Such line-ups are a defining feature of orgax in which the contributing factors arise at many levels of the system – the workplace, the organization and the regulatory environment – and subsequently combine in often unforeseen and unforeseeable ways to allow the occurrence of an adverse event. In well-defended systems, such as commercial aircraft and nuclear power plants, such concatenations are very rare. This is not always the case in healthcare, where those in direct contact with patients are the last people to be able to thwart an accident sequence.

Latent conditions possess two important properties: first, their effects are usually longer lasting than those created by active failures; and, second, they are present within the system prior to an adverse event and can – in theory at least – be detected and repaired before they cause harm. As such, they represent a suitable target for safety management. But prior detection is no easy thing because it is very difficult to foresee all the subtle ways in which latent conditions can combine to produce an accident.

It is very rare for unsafe acts alone to cause such an accident – where this appears to be the case, there is almost always a systemic causal history. An obvious domain where unsafe acts might be the sole factor is healthcare – where the carer appears to be the last line of defence. Three healthcare case studies are among the 10 discussed below. In each, the unsafe actions of the immediate carers are shaped, even provoked, by systemic factors.

Promising candidates for close study are the generic organizational processes that exist in all systems regardless of domain – designing, building, operating, managing, maintaining, scheduling, budgeting, communicating and the like.

This book extends and develops these ideas using case studies that have occurred in a variety of domains in the period that has passed since the 1997 book was written and published. These analyses provide the 'raw data' for the process of drilling down into the underlying causal pathways. Many contributing latent conditions recur in a variety of domains. A number of these – organizational issues, design, procedures and communications in particular – are examined in detail in order to reveal likely problems before they combine to penetrate the defences-in-depth.

Beyond this point, the book is divided into two parts. The first, comprising Chapters 2 and 3, summarizes the basic arguments underlying orgax and the unsafe acts (or active failures) that can contribute to them. These are intended as starters for those who haven't read the 1997 book or have forgotten it.

The second and main part of this book contains seven chapters that go beyond the mid-1990s. Chapter 4 digs down into the factors underlying latent conditions. Promising candidates are the generic organizational processes that exist in all systems regardless of domain – designing, building, operating, managing, maintaining, scheduling, budgeting, communicating and the like.

Chapter 5 extends and develops these ideas using a series of 10 orgax case studies that have occurred in a variety of domains in the nearly 20 years that have passed since the first book was written and published. Three are taken from healthcare, two involving the unwanted release of radiation, one railway accident, two explosions of hydrocarbons and two aviation accidents. They show the almost unimaginable ways in which the contributing factors can arise and combine. These analyses provide the 'raw data' for the process of drilling down into the underlying causal pathways. Many contributing latent conditions recur in a variety of domains.

Chapter 6 discusses a well-publicised regulatory disaster relating to an NHS Foundation Trust. The various layers of regulation failed to identify a very distressing number of shortcomings in this hospital. It is highly likely that these failures are not unique to this hospital. Even as I write, news is breaking of a comparable set of regulatory deficiencies in a maternity hospital in the north-west of the UK.

Chapter 7 describes foresight training: a set of measures that are designed to make people at the sharp end more 'error-wise' and aware of the situational risks. These often form the last and all-too-neglected line of defence. It addresses the issue of what mental skills can we give sharp-enders to make them more alert to the dangers.

Chapter 8 looks at alternative theoretical views. These are important because it is often assumed that the Swiss cheese model is the principal explanatory metaphor. But, as you will see, it has its critics – and rightly so.

Chapter 9 is mainly concerned with patient safety. It traces a cyclical patient journey that looks both to the past and to the foreseeable future.

Chapter 10 asks the following question: is any kind of optimism justified in the matter of organizational accidents? The answer offered is a very tentative maybe.

Chapter 11 relates two stories of heroic recovery: the 2009 'miracle on the Hudson' and the saving of the Japanese Fukushima Daini nuclear reactor in 2011.

PART 1
Refreshers

Chapter 2
The 'Anatomy' of an Organizational Accident

As outlined in Chapter 1, orgax occur in complex systems possessing a wide variety of technical and procedural safeguards.[1] They arise from the insidious accumulation of delayed-action failures lying mainly in the managerial and organizational spheres. Such latent conditions (or latent failures) are like resident pathogens within the system. Organizational accidents can result when these latent conditions combine with active failures (errors or violations at the 'sharp end') and local triggering factors to breach or bypass the system defences. In this chapter and Chapter 3, we will examine in some detail the processes that give rise to latent condition. For now, I should stress that defensive weaknesses are not usually gaping holes or absences; rather, they are more like small cracks in a wall.

Figure 2.1 sets out the basic 'anatomy' of an organizational accident. This model is now used in a variety of hazardous technologies to guide accident investigations and to monitor the effectiveness of remedial measures. The direction of causality is from left to right:[2]

- The accident sequence begins with the negative consequences of organizational processes (i.e., decisions concerned with planning, forecasting, designing, managing, communicating, budgeting, monitoring, auditing and the like). Another very influential factor is the system's safety culture.

---

1   Reason, J. (1997). *Managing the Risks of Organizational Accidents*. Aldershot: Ashgate.
2   Reason, J. (2008). *The Human Contribution*. Aldershot: Ashgate. See Chapter 5 for the various diagrammatic representations of the Swiss cheese model.

- Some of the latent conditions thus created are transmitted along departmental and organizational pathways to the various workplaces where they show themselves as conditions that promote errors and violations (e.g., high workload, time pressures, inadequate skills and experience and poor equipment)
- At the level of the individual at the 'sharp end', these local latent conditions combine with psychological error and violation tendencies to create unsafe acts. Many unsafe acts will be committed, but only very few of them will penetrate the many defences and safeguards to produce bad outcomes.
- The fact that engineered safety features, standards, administrative controls, procedures and the like can be deficient due to latent conditions as well as active failures is shown by the arrow connecting the organizational processes to the defences.

It is clear from the case studies discussed above that the people at the human–system interface – the 'sharp end' – were not so much the instigators of the accident; rather, they were the inheritors of 'accidents-in-waiting'. Their working environments had been unwittingly 'booby-trapped' by system problems.

When systems have many layers of defences, they are largely proof against single failures, either human or technical. The only

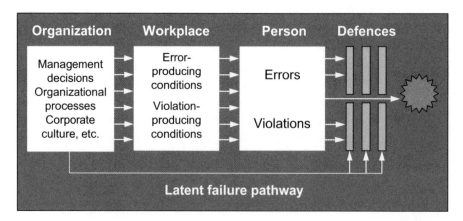

Figure 2.1   Summarizing the stages involved in an organizational accident

types of accidents they can suffer are organizational accidents, that is, ones involving the unlikely combination of several factors at many different levels of the system that penetrate the various controls, barriers and safeguards (see Figure 2.1 on the previous page for a diagrammatic representation – it does not replace the traditional SCM picture, but is instead laid out to emphasize the sequence of causality).

In Chapter 5, we look in detail at case studies describing 10 orgax, none of which were discussed in the first edition. The aim here is to describe the many contributing factors and the unlikely conjunctions that constitute actual orgax.

Chapter 3
Error-Enforcing Conditions

The principal sources of latent conditions in the 10 case studies described in Chapter 5 are summarized in Table 3.1. Before considering these in more detail (see Chapter 4), it is important to note that all of the case study accidents involved active failures both at the sharp end and through various levels of the system – this is not always the case (e.g., the King's Cross Underground fire). Latent conditions alone can sometimes be sufficient to breach the defences.

Table 3.1 **The principal sources of latent conditions related to specific case studies**

Case 1	Case 2	Case 3	Case 4	Case 5
Organization	Communication	Procedures	Organization	Goal conflict
Communication	Hardware	Design	Procedures	Hardware
Hardware	Procedures	Organization	Hardware	Procedures
Design	Design	Hardware	Supervision	Organization
				Culture
				Training

Case 6	Case 7	Case 8	Case 9	Case 10
Communication	Hardware	Goal conflict	Goal conflict	Training
Training	Context	Hardware	Organisation	Design
Organization	Design	Procedures	Supervision	Hardware
Procedures		Organization	Procedures	Communication
			Culture	
			Training	

Cases: 1. Anaesthetic fatality; 2. Glenbrook rail crash; 3. Omnitron; 4. Engine Oil; 5. Tokai; 6. Vincristine; 7. GP; 8. Texas City; 9. Deepwater Horizon; 10. AF 447

Active Failures (Unsafe Acts): A Brief Refresher

Unsafe acts are of two distinct types: *errors* and *violations*. Errors arise from informational problems and fall into three categories: *skill-based slips* and *lapses, rule-based mistakes* and *knowledge-based mistakes*. Violations arise from motivational factors and fall into four types: *routine (or corner-cutting) violations, thrill-seeking or optimizing violations, necessary violations* and *exceptional violations*.

Everyone makes errors. They are an intrinsic part of the human condition, but certain situations are guaranteed to increase the chances that errors will be made. These are called (here) error-enforcing conditions. When they occur in the presence of dangers, there is a heightened risk of injury or damage.

Slips and Lapses

These are actions-not-as-planned. Absent-minded slips arise from inattention. Lapses are failures of memory – when we forget to carry out an intended action or forget that we have already performed a particular action.

Mistakes

Here the actions go as planned, but the plan itself is inadequate to achieve the desired goal. Mistakes occur when we are trying to solve problems or are dealing with unexpected departure from routine. There are two kinds of mistakes: rule-based, when we are dealing with a familiar problem but apply the wrong rule, or knowledge-based, in which we have run out of ready-made solutions and have to think on our feet in novel circumstances.

Conditions Promoting Errors

Slips and lapses are most likely to happen when we are performing habitual actions in familiar surroundings. Almost invariably, our minds are on something other than the task in hand. People are very good at performing routine tasks so long as things remain unchanged.

Two conditions in particular provoke slips – one psychological and the other situational:

- The psychological condition is attentional capture. This where an individual is so distracted or preoccupied with something other than the task in hand that they have little or no attentional capacity to monitor the progress of their current actions.
- The situational condition is change. This can be either a change in the nature of the task or in the circumstances in which the task is performed. The most error-enforcing changes are those in which there is only a partial alteration in the circumstances of a familiar task. In other words, most features of the task remain the same, but one or two significant aspects are changed. The worst situation is where these changes demand actions that are quite different from those normally required.

Given these two conditions, the nature of the slip is predictable – people will do what they customarily do in those circumstances, having failed to attend to the features that require them to modify their actions. Errors tend to take forms that are more familiar, more frequent-in-context than those that were either intended or required. If you want to know what kind of error a person will make in a particular changed circumstance, then consider what he or she did before the change was made. In summary, a very common kind of slip, perhaps the commonest, is a strong-habit-intrusion.

Some of the mental factors associated with making mistakes are different from those for slips, while others are very similar. One important difference between mistakes and slips is that in the former case, we are usually aware that we have a problem or some departure from the expected. In short, we are operating in a conscious rather than an automatic mode.

Rule-based mistakes, however, usually share with slips the fact that the error takes a frequent and well-used course of action, especially if they are solutions to familiar problems and have served us well in the past. It is only natural that they spring readily to mind when we encounter a problem that has many features in common with those that we have frequently met before. These similarities between problem solutions cause us to overlook important differences. Driving, for example, is a highly routinized activity, performed very largely on automatic pilot. If we fail to notice that the road conditions have become slippery, we are likely to hit the brakes in an emergency rather than steering out of trouble.

Rule-based mistakes come in two basic forms: either we can misapply a normally good rule or we can apply a bad rule. In learning a job, many people acquire bad habits or bad rules. Often these bad habits go unpunished and become a fixed part of a person's skill repertoire. But sometimes the circumstances are less tolerant. Consider the case of the technician rewiring a signal box on the day before the Clapham rail disaster. He had the practice of bending wires back rather than removing them. In this instance, the bent wires uncoiled and made contact, causing the signal to fail green, allowing a commuter train (the next morning) to crash into the back of a stationary train. The debris derailed a train coming in the opposite direction.

In the case of knowledge-based mistakes we have exhausted our stock of pre-packaged solutions and have to tackle a novel situation by resorting to the highly fallible of business of thinking 'on the hoof'. Although people can be quite good at coming up with fresh solutions when they have plenty of time to reflect, they generally perform badly under the more usual conditions of time pressure, strong emotion and the prospect of imminent disaster (e.g., Air France Flight 447). In this situation we are forced to make progress through trial-and-error. The military have always known this. This is why soldiers are drilled to carry out the necessary emergency actions when their weapons fail in the heat of battle.

There are at least three reasons why people perform badly when trying to solve novel problems:

1. Human attention span (what is available to the conscious mind at any one time) is usually much smaller than the full scope of the problem. We only have a keyhole view of the total picture. This means that we have to shift our attention

over the full range of the 'problem space'. As we do this, information spills out of short-term memory and we forget what we have just done or thought about.

2. When we encounter a totally new situation, we have to resort to a 'mental model' (a picture in the mind's eye). This is almost always incomplete or inaccurate, or both.

3. The two items above relate to the way we process information. The third reason has to do with emotions. Stress, the awareness of danger, will act to cloud thought. In particular, it causes our actions to become more stereotyped. We very easily give up the effortful business of thinking and resort to a familiar and comforting sequence of habitual actions.

As we shall see later, situations that demand novel thinking, especially when a great deal depends on getting it right, are very productive of error. Indeed, making errors is one way of steering out of trouble. Errors mark the boundaries of an acceptable way forward.

Error Probabilities

Assigning probability values to errors is still very much an art form.[1] Nonetheless, there is sufficient general agreement about the likelihood of errors occurring in various generic tasks to put them into a sensible rank order. Some of these tasks are listed in Table 3.1 at the start of the chapter. Their main use is to provide bench marks for the range of error probabilities over different jobs, see Table 3.2 on the next page.

Table 3.3 lists a number of well-researched error-enforcing conditions (EEC). For each one, there are the following:

- *Description*: a general description.
- *Risk factor*: a number (x10) representing the approximate amount by which the nominal error probability will increase when someone encounters this EEC initially (the error probability cannot exceed 1.00).

---

1   SIPM-EPO/6. *Tripod. Vol II: General Failure Types*. The Hague: Shell International Petroleum Company.

- *Error type*: the most likely types where these can be predicted on psychological grounds. The conditions are listed in order of severity, beginning with the most provocative conditions.

Table 3.2   Error-enforcing conditions and their relative effects

Generic Task	Nominal Error Probability
1. Carrying out a totally novel task with no real idea of the likely consequences.	0.75 (0.50–1.00)
2. Attempting to repair a fairly complex system on a single attempt without direct procedures or the procedures to hand.	0.30 (0.20–0.30)
3. Performing a complex task requiring a high degree of knowledge and skill.	0.20 (0.10–0.35)
4. Performing a moderately simple task quickly and with scant attention.	0.10 (0.06–0.28
5. Restoring a piece of fairly complex equipment to its original state following maintenance.	0.007 (0.0008–0.010)
6. A highly familiar routine task performed by a well-motivated and competent workforce.	0.0005 (0.00008–0.0010)

Table 3.3   Error-producing situations

Category	Context	Risk Ractor	Type
Unfamiliarity	A novel and potentially dangerous situation	17	KB mistake
Time shortage	Time available for error detection or correction	11	All types
Noisy signals	Low signal to noise ratio in communications between groups	10	Misperception
Poor interface	Equipment fails to communicate vital signals or conflicting controls	8	Action slips
Designer/user mismatch	Equipment conflicts with user's model of the world	8	All types
Irreversibility	Unforgiving equipment item can't correct actions	8	All types
Overload	Too much information badly presented	6	Perceptual slips
Technique unlearning	Need to apply new technique having reversed action	6	Strong habit intrusions

Table 3.3   Continued

Category	Context	Risk Ractor	Type
Knowledge transfer	Need to transfer knowledge from task to task without loss	5	Strong habit intrusions
Misperception of risk	Mismatch between real and perceived risk	4	Mistakes
Inexperience	Training or inexperience insufficient	3	Mistakes
Checking	Little or no checking of job done	3	Undetected errors
Culture	Macho culture, dangerous procedures	2	Accident
Morale	Morale low, loss of faith in leaders	1.2	Sloppiness
Monotony	Boredom, repetitive tasks	1.1	Unvigilant
External pacing	Pressure of supervisors	1.06	Slips

PART 2
Additions since 1997

Chapter 4
Safety Management Systems

There are very few managers of hazardous technologies who are unfamiliar with Safety Management Systems (SMSs). They are products of a steady series of moves towards self-regulation, or even deregulation – part of a regulatory progress that began in the 1970s and became a standard feature of the 1990s and beyond. Now they are virtually universal. In brief, the progress was a gradual transition from reactive to proactive methods of managing and controlling hazards. A detailed history of this development is given in *Managing the Risks of Organizational Accidents*,[1] but I will outline the main stages below:

- In 1974, Lord Robens headed the inquiry into the Flixborough disaster. The recommendations of the Robens Committee formed the basis – with remarkably little governmental tinkering – of the Health and Safety at Work Act (HSW). This was the springboard for the current state of proactive regulation. Unlike the Factories Act that preceded it, HSW did not go into great detail with regard to specific accident producers.
- HSW gave rise to two new bodies: the Health and Safety Commission (HSC), which was responsible for achieving the general purposes of the Act, and the Health and Safety Executive (HSE), which was charged with the oversight of health and safety policy.
- In 1984, the HSE issued the Control of Industrial Major Hazards (CIMAH) regulations. This required that the operator should provide the HSE with a written report on

---

1   Reason, J. (1997). *Managing the Risks of Organizational Accidents*. Aldershot: Ashgate, pp. 175–82.

the safety of the installation, known as the Safety Case. Once accepted, the HSE used it as a basis for its inspection strategy. Considerable importance was attached by the HSE to management and organizational issues.

- In 1988, the Control of Substances Hazardous to Health (COSHH) Regulations became law in response to a European Directive on Hazardous Agents. This set out the guidelines for what was needed in order to protect employees and others. It went further than HSW in requiring employers to be aware of the properties of over 5,000 substances. It also laid down a 'rolling procedure' for the identification, assessment and control of risk.

- The next major development followed directly from the recommendations set out in Volume 2 of the Cullen Report on the Piper Alpha disaster, which was published in 1990. This was the 'formal safety assessment' that must be performed by each hazardous installation – also known as the Safety Case. This needed to be updated at regular intervals and on the occurrence of a major change of circumstances. This became the SMS.

Is the SMS regime an improvement on what went before? It is in one very important respect. Their predecessor, the Factory Act, focused upon specifics – the number and placement of fire extinguishers, machinery, hoists and lifts, etc. – things that for the most part targeted the prevention of individual accidents. However, the SMS focuses squarely on those things – organizational structure, management, etc. – that are implicated in the occurrence of organizational accidents. It does not ignore individual accidents, but they are not centre stage. Has this helped? Judging by the number of orgax that have occurred during the SMS regime, probably not – though it has surely made things better, particularly in the way that it has involved management in the safety process. Another huge bonus is that the creation of formal safety assessments has made system operators become more aware of the hazards that confront them, often for the first time. But there are problems.

Between a Rock and a Hard Place

The Factories Act sets forth very specific requirements. These gave the regulator very clear indications of operator compliance or otherwise, but SMS rarely does. The regulator has two very clear requirements: first, to assess the acceptability of the written Safety Case; and, second, to determine the operator's compliance to that Safety Case.

Following a major accident, there are two possibilities: the operator was compliant with the Safety Case – in which case, the Safety Case was flawed and should not have been approved; or there was non-compliance – indicating failures of regulatory oversight. Either way, the regulatory body has a big problem.

Beyond the Ring Binders

Some years ago, I visited a very prestigious national body and asked the safety manager to outline his safety-related measures. He took me into his office and pointed up to a wall-to-wall shelf that was filled with fat ring binders. 'That's our Safety Case' he told me. It was evident that these thousands of words had occupied a large number of man hours. The issue that concerned me was the extent to which these ring binders corresponded to the situations that existed within the organization's many distributed workplaces. No one knew for sure – although the organization in question had an exemplary safety record, having very few safety significant incidents.

Evaluating Safety Cases

How does a regulatory body judge whether a particular written Safety Case is fit for purpose? There is likely to be little or no uniformity in individual SMSs. Each one is tailored for a particular purpose in a wide range of hazardous domains. Another problem is that most regulators have a technical background; there are relatively few likely to be well versed in such 'soft' issues as human factors, man–machine interactions or organizational theory. Predicting the aetiology of an organizational accident is very difficult, even for the so-called experts. Of course, regulators can call upon experience, but this is likely to be limited and domain-specific. I think many regulators would agree that it's an almost impossible task and yearn for the days of hard-and-fast standards. Given that a large part of their job is evaluating written material, there must be a great temptation to weigh it rather than read it.

Chapter 5
Resident Pathogens

I have often referred to latent conditions as 'resident pathogens'. A resident pathogen is something that may lurk within the body for some time before it shows itself as an illness. Glandular fever, for example, may seed pathogens into the body without immediate consequences, as can many other conditions.

So the starting questions in this chapter are: what are the systemic equivalents in complex hazardous technologies? What factor is guaranteed to create resident pathogens? The answer, I believe, is the inevitable and universal conflict between production and protection or – more directly – between profit and safety. In a phrase: cost-cutting.

Cost-Cutting

Clearly, there is nothing intrinsically wrong with keeping costs down to a workable minimum. This is what managers are appraised for. The production versus protection quandary is not easy to resolve: both need to be commensurate with one another. Too much production at the cost of protection spells possible disaster; too much protection without production to pay for it risks bankruptcy. There are two major principles:

- The ALARP principle: keep your risks 'as low as reasonably possible'.
- The ASSIB principle: 'and still stay in business'.

Navigating between these two imperatives is difficult. The first problem is the disparity in the quality of feedback relating to these two goals. Information relating to production is continuous, compelling and relatively accurate, but that concerning protection (safety) can be sparse, intermittent, difficult and often untrue. For example, many large organizations (if not most) working in hazardous domains base their assessments of system safety on the rates of individual accidents (lost time injuries (LTIs)), but as discussed at length in the first book and elsewhere, individual accident rates do not predict the likelihood of an orgax – as was conceded in the BP report on the Texas City explosions.[1] The road to hell is paved with falling LTI frequency rates.

The managers of hazardous technologies tend to be very numerate people. Their instincts and their training favour numbers. And the lost time incident frequency rate serves that end very well. They allow comparisons between companies and departments; they appear to provide an ongoing measure of relative safety; as such, they are just what an MBA-trained manager ordered. But they are a procrustean trap when it comes to predicting orgax – resulting from the unforeseen and often unforeseeable concatenation of causal factors arising from many parts of the system. The presence of defences-in-depth fortunately makes such conjunctions rare events, so they are hardly ever uppermost in the minds of those primarily pursuing productive goals. It is therefore tempting to think that if we do today what we did yesterday, nothing bad will happen – and for the most part that will be true. So what can managers do for the best?

I don't believe that there are any measures that will guarantee complete freedom from some future orgax as long as the working conditions continue to be hazardous. But there are things that managers and system designers can do that will stand a fair chance of delaying that evil day. In answer to the question posed above, managers can take steps to improve a system's safety health.

How do we assess health? We can take a leaf out of the doctor's book and sample a subset of the body's physiological processes: blood pressure, heart rate, urine content, blood content analysis and so on. We don't have to measure everything: most things are connected. Table 3.1 gave us several candidate subsystems that are implicated in the development of an orgax. Most commonly (at least in our selected case studies), they are organization, hardware,

---

1   BP Fatal Accident Investigation Report, Texas City, Texas, USA, 9 December 2005.

procedures, supervision and design – and, of course, culture, which is a property of all of these indices. But this has been discussed at length elsewhere. The purpose of the remainder of this chapter is to look more closely at a limited sample of these indices.

Let me pause the narrative at this point and go back to a day in the late 1980s. I received a call from Professors Willem Albert Wagenaar and Patrick Hudson at Leiden (the Oxbridge of the Netherlands). They said that they had just been offered a large amount of money from Shell (an Anglo-Dutch oil and gas company based in The Hague) and would I help them spend it. It was a characteristically generous offer which I could not refuse.

For the next five or six years, we travelled around the world looking at safety in Shell's operating companies and its large tanker fleet. As a company, it was still hooked on LTIs, but was well aware that dangerous processes needed other ways of gauging safety.

Our contribution was to devise various kinds of proactive measurement of what we termed General Failure Types (GFTs): these were hardware, design, maintenance management, procedures, error-enforcing conditions, housekeeping, incompatible goals, communication, organization, training and defences. I am not sure I would now wish to retain them all: they are best suited to oil and gas exploration and production. But at least six of them fit well with our current need for generic systemic processes that are more or less universal.

Then we developed a tool that was called Tripod-Delta.[2] Tripod-Delta had three fundamental elements:

- A coherent safety philosophy that allows the setting of attainable safety goals.
- An integrated way of thinking about what disrupts safe operations.
- A set of instruments for measuring GFTs that do not depend on outcome measures.

We can summarize the underlying philosophy as follows. Safety management is essentially an organizational control problem. The

---

2   This is discussed in detail in Reason, J. (1997). *Managing the Risks of Organizational Accidents*. Aldershot: Ashgate, pp. 131–8.

trick is to know what is controllable and what is not. Bad outcomes arise from the interaction between GFTs and local triggering factors, but only the GFTs are knowable in advance. The important thing about Tripod-Delta is that it is a bottom-up device, that is, the people at or close to the sharp end tell management how the world really is rather than how it ought to be.

For the remainder of this chapter, we will look in some detail at what can go wrong with four GFTs: organization, hardware, procedures and design. All of them have featured in the ten case studies discussed in the next chapter.

Organization

Organizational failures are deficiencies in either the structure of a company or the way in which it conducts its business that allow safety responsibilities to become ill-defined and warning signs to be overlooked. Certain aspects of safety get lost in the organizational cracks.

Perhaps the most dramatic example of this is London Underground Ltd just prior to the King's Cross Underground fire in November 1987. A burning cigarette fell through a crack in an old wooden escalator and ignited oily debris; the result was that 31 people died. This escalator seems to have disappeared from the London Underground organogram. One consequence was that the escalator received inadequate maintenance.

London Underground shared with other railway organizations a long tradition of ensuring passenger safety. However, this tradition focused, naturally enough, upon the mitigation of obvious risks involved in actually travelling on the world's largest and oldest underground railway. Just as the lit cigarette fell through a crack in a 50-year-old wooden escalator on to long-established layers of flammable debris, so did the responsibility for the safety of passengers in stations drop between large gaps in the London Underground organogram.

Following London Underground's reorganization in the early 1980s, responsibility for safety was scattered across three of its five directorates. No one part had overall safety responsibility. No one was in a position to view the total safety picture. To complicate matters further, responsibility for escalator and lift management had recently been contracted out. The task of checking on this work was assigned to a junior member of the Engineering Directorate who was based some distance from King's Cross Station.

At that period in its history, London Underground's primary concern was to make the railway profitable. Rationalization, cost-cutting, leaning and meaning were the orders of the day. Safety was given over to the departmental specialists and the regulators. Both the safety professionals and the senior management regarded the frequent fires ('smoulderings') on Underground stations as being inevitable in such an elderly, large and under-capitalised railway network.

This unwarranted insouciance and normalization of deviance followed a similar pattern to that in the period prior to NASA's Challenger disaster. NASA and Morton Thiokol were well aware of the frequent scorching on the shuttle's O-rings for several years prior to the accident, just as London Underground had a long history of station smoulderings. The political and commercial pressures were such as to cause senior management to underplay these now obvious threats.

The moral of these and other case studies is that a combination of incompatible goals and organizational shortcomings can cause clear warning signs to be disregarded, either through wishful thinking or because the company is badly structured so that no one has the unambiguous responsibility for doing something about them.

Shell, I believe, has addressed these problems very effectively by making safety a line management responsibility. This has a number of important organizational consequences, which are listed below:

- Safety departments should act in an advisory capacity. They are neither responsible nor accountable for safety performance.
- It is incumbent on senior management to develop safety policies and to create the safety organization necessary to implement these policies.
- All line safety responsibilities should be clearly defined in job descriptions.

- The effectiveness of each manager's safety attitudes and performance should be assessed as part of the staff appraisal procedure.
- There should be organizational mechanisms for establishing safe standards and procedures for both operations and equipment.
- All employees should be made aware of their individual safety responsibilities.
- A two-way flow of safety information and ideas should be encouraged.

Ron Westrum, an American social scientist, identified three ways in which organizations habitually deal with safety problems:

- *Denial* – problems are not admitted and whistleblowers are discredited or dismissed.
- *Repair* – problems are admitted, but they are fixed at a local level and their wider implications are ignored.
- *Reform* – the problem is seen as having its origins beyond the immediate circumstances and broadly based action is taken to reform the organization as a whole.

There are no easy solutions to organizational problems. Healthy organizations are characterized by their commitment to continuous self-assessment and reform. Effective reform actions require two preconditions. The first is a willingness to take the long view on safety and to tackle underlying health problems rather than the short-lived symptoms. The second is a preparedness to endure a state of chronic unease whenever safety matters are considered. In the safety war, clear successes are few and far between. What is needed is the will to wrestle continuously with GFTs. It is highly unlikely when all of these failure types are non-existent. Fighting the safety war is 'one damn thing after another'.

Hardware

Hardware deals with the quality and availability of tools and equipment. Equipment is used here in the broadest sense possible: a condensate pump is as much a piece of equipment as the whole platform on which it is installed. Tools are the instruments used to operate, maintain and repair equipment.

In comparison with some of the other GFTs, the quality of the hardware is more tangible and more readily discovered. Hardware is something that most technical managers understand and are comfortable with. Problems appear in one of three phases: the operation phase, the construction phase and the availability phase.

Most problems arise from either cost-cutting or confusion: differences of opinion about specifications, cheap brand purchasing, buying the wrong equipment and failure to renew.

The symptoms of hardware problems are listed below:

- Re-ordering of the same piece of equipment.
- Frequent manufacturer call-back.
- High incidence of breakdowns.
- More hours of maintenance than expected.
- High number of ordered spare parts.
- Unused or abandoned equipment.

The availability of equipment may be affected by either or both of the following:

- Poor ordering system.
- Poor logging system.

Procedures

Procedures are so commonplace in complex hi-tech industries that their nature and purpose tend to get taken for granted. It is worth starting this section by reminding ourselves of what procedures are and why they are necessary.

Procedures communicate task know-how. They become a GFT when they are unclear, unavailable or otherwise unusable. Collectively, they comprise the accumulated craft wisdom and practical knowledge of the company as a whole. Good procedures should tell people the most efficient and safest way of carrying out a particular task. It should be noted that clumsy procedures are among the main causes of human violations. Good procedures are necessary for three reasons:

1. Many jobs are too complicated for their individual steps to be self-evident.
2. The information necessary to perform tasks correctly is usually too much to be carried reliably in human memory.
3. People change faster than jobs. This means that the organization needs a means of passing on standardized task information from one person to the next.

In assessing procedures, we need to take account of the following factors:

- Are the procedures accessible to those who use them?
- Are the procedures intelligible and unambiguous?
- Do the procedures promote omissions?
- Do the procedures contain omissions or errors?
- When did the procedure writers last attempt to execute their own procedure?
- Do the procedures contain enough contingencies to allow for an imperfect world?
- What is the 'between the lines' safety message being communicated?

Design

Design can become a GFT when it affords particular types of errors and violations. Design has been defined as the optimum solution to the sum of the true needs of a particular set of circumstances. In reality, most designs fall short of being optimal solutions. In the first place, this is especially true of the needs and capabilities of the human users of the system. This lack of human factors knowledge has been discussed at length in the first edition of this book. In the second place, designers are hedged around by a large number of constraints. Thus, the final design is inevitably a compromise between the wishes expressed in the clients' technical specification and the necessity of accommodating the various constraints imposed by the human users.

What exactly is meant by design-induced affordances for errors and violations? Two examples will illustrate the point:

- Most (now outdated) digital cameras are fitted with electronic devices that remove from their users the need to understand the subtleties of shutter speeds and aperture characteristics. The fact that these cameras were designed to be idiot-proof clearly reveals some awareness of human frailty on the part of the designers. But, in many models, this has afforded another possibility of error – that of forgetting to switch off the camera before putting it away in the carrying case, thus running down the batteries unnecessarily. This on/off switch is often located on the top of the camera and its operation is easily omitted. Had the designers taken this knowledge into account, they would have introduced a forcing function, that is, they could have made it impossible to put the camera back in its case until it had been switched off. This is easily achieved by locating the on/off switch beneath the wind-on lever. Many cameras are designed so that they can only be replaced in the carrying case once the wind-on lever is flush with the moulded body of the camera. This same action could also serve to turn off the power.

- Affordances for violations often involve designs that require users to follow something other than a direct path to their immediate goal or destination. This is often seen in the right-angled cornering of paths within urban parks and other open spaces. Here the designer has been working to satisfy some aesthetic criterion. His or her neglect of the fact that people will always take short-cuts whenever opportunities present themselves is revealed in the unofficial muddy diagonal paths cutting through the grass or flowerbeds. The designer's aim was to give the user a pleasant visual experience; what he or she overlooked was that the open space also affords access to some more pressing goal, like a bus stop, rail station or shopping centre.

The Separation of Design and Construction

Before the Industrial Revolution, the tasks of designing and making were usually part of the same process. For example, the shape of cartwheels for horse-drawn vehicles slowly evolved over the centuries into an elaborate dish-shaped geometry.

Craft-based designs of this kind were possible when objects, like cartwheels, muskets and wooden battleships, operated on much the same basic principles over many centuries. With each succeeding generation, these objects gradually approximated to their optimal forms within the limits of the available technology.

The need to separate and professionalize the design and construction functions came when technologies began to develop and change at an extremely rapid rate. Design by drawing necessarily superseded design by making. Drawing boards and computer-aided design (CAD) systems offer many mental freedoms not available to the material-bound craftsman, but the architects of high-rise housing Britain in the 1960s could not easily visualize the social problems associated with these new forms of public housing.

A general problem for modern professional designers, then, is that they have become increasingly remote – professionally, physically and in time – from the flesh-and-blood realities in which their designs will function. And this separation is most evident in their frequent lack of awareness of the capabilities and limitations of the end-user – someone who may belong to a different culture, having only modest educational attainments, and who is likely to be working in far from ideal physical conditions.

Discrepancies between Designers' and Users' Mental Models

These discrepancies have been described most eloquently by Donald Norman, a distinguished American cognitive psychologist.[3] Such gulfs are of three basic kinds.

1. *The knowledge gulf*
   There are two kinds of knowledge: *knowledge-in-the-head* (KIH) and *knowledge-in-the-world* (KIW).

   KIH is contained in memory and is thus largely independent of the immediate surroundings, but it needs to be learned and retrieved. To acquire it requires a major investment of time and effort. To use it in a timely way often requires reminding.

---

3   Norman, D.A. (1988). *The Psychology of Everyday Things*. New York: Basic Books.

KIW, on the other hand, is located in the external world rather than in memory. These are the cues which lead us naturally and automatically from one action to the next, or the forcing functions that physically bar unwanted progress along unwanted or unsafe pathways. KIW is always there and does not need prompting, but it is subject to the out-of-sight-out-of-mind principle. KIW does not have to be learned, merely acquired and interpreted. Ideally it does not have to be thought about, it is simply acted upon. It supports the kind of skill-based performance at which human beings excel.

Designers, sitting at their drawing boards, of necessity rely heavily on KIH. As such, they frequently fail to appreciate how much the end-user depends upon KIW to achieve the error-free operation of the designed item. Many designed-induced errors arise because designers underestimate the extent to which necessary knowledge should be located in the world rather than in the user's head.

2. *The execution gulf*

Designed objects or equipment are often opaque with regard to their inner workings or to the range of possible actions that are safe. Nor do they always reveal what the user must do to achieve a particular goal. Thus, they fail to afford the actions necessary for correct and/or safe performance.

Doors are a classic example. On approaching a set of double doors, the user needs to know which side opens and whether it should be pushed or pulled. Good design puts this knowledge in the world. Visible signals are provided for the correct operation. The handle design tells the user both where and how to act. Some handles afford only pushing – there are no grasping points to allow pulling. Pulling cues are more difficult since one that allows pulling also permits pushing. In this case, it is necessary to exploit both forcing functions and cultural stereotypes to ensure the appropriate action.

3. *The evaluation gulf*

Equally as important as directing the user to the appropriate action, a designed item must also afford users clear and unambiguous knowledge of the current state. Dangerous errors occur because users are unable to gauge the consequences of their actions accurately. Pilots of the earlier versions of the Airbus aircraft were frequently heard to say: 'What's it doing now?'

A major contributor to the near-disaster in 1979 at the Three Mile Island nuclear power station was the design of the control panel. The operator pushed a button to close a valve, but the valve failed. Nevertheless, the light on the panel clearly indicated a closed valve. The light did not actually monitor the position of the valve, only the actuating signal from the button to the valve. The panel indications failed to distinguish between the valve being commanded shut and actually being shut. The problem here lay not so much with the operators as with the design of the panel.

Design Principles to Minimise Error Affordances

Awareness of these possible discrepancies between the mental and users models of designers leads to a set of design principles that should reduce error affordances. We start with some basic questions that need to be asked about the designed item:

- How easy is it for the user to work out what the item does?
- Is the range of possible actions readily apparent to the user? Are these features directly visible?
- Is it immediately obvious to the user what they need to do to achieve some particular goal?
- Is it easy for them to actually perform these actions?
- Having performed these actions, how easy is it for the user to find out whether the item is in the desired state? Is the state of the system readily apparent?

If the answers to these questions are in the negative, the designers might respond by saying that the necessary instructions are available in the instruction book. Clearly some pieces of equipment are so complex that they require the accompaniment of written instructions. But it is a good design principle to keep this written knowledge to a minimum and to convert as much of this necessary guidance into visible knowledge of the world. There are many reasons for this.

People by and large are disinclined to read instruction manuals carefully. This is not just laziness, though this can play a part. Often, instruction manuals are badly written (or translated), not to hand when needed, or the work is being done in conditions that make reading difficult or impossible.

In any case, people prefer to fiddle with things rather than read about them. In this, they are being true to their psychological nature, which prefers action to thought and pattern-matching to detailed analysis. As has been stated many times, humans function most comfortably at the skill-based level of performance. They are, of course, capable of knowledge-based functioning, but it is slow, error-prone and laborious. It runs counter to the basic mental principle of least effort. These are not universal personality defects, they are simply a reflection of the strengths and weaknesses of human mental functioning. It is the way we evolved.

Conclusion

In the next chapter, we examine 10 case studies of organizational accidents. On the surface, they are very different, involving different systems in different circumstances. But underlying these disparate surface features, there are some uncanny similarities that are the hallmark of orgax. They all involve failed defences and the diabolical concatenation of many pathogens arising from different parts and varying levels of the system – just the kinds of events that are unlikely to be predicted by even the most sophisticated SMS.

Chapter 6
Ten Case Studies of Organizational Accidents

We have considered the general characteristics of orgax. Now it is time to get down to the nitty-gritty. This chapter presents case studies of orgax from a wide variety of domains. They are presented in some detail. It is highly unlikely that any one reader will be familiar with all of the case study areas. But a detailed knowledge of the technicalities within each domain is not necessary. I hope you will acquire some understanding of the pathways and the often unforeseen and unforeseeable manner in which the contributing factors combine to cause disaster. Although we shall be dealing with the minutiae, the need here is to keep your eyes upon the bigger picture.

Case Study 1: An Anaesthetic Fatality[1]

The possibility of vomiting under anaesthesia – carrying the risk of death through aspiration of the vomitus – is a well-known surgical hazard. It is normally defended against by pre-operative fasting, by vigilant nursing, and by a detailed assessment of the patient by the anaesthetist both on the ward and during the pre-operative preparation. In this instance, a combination of a number of both active failures and latent conditions contrived to breach these 'defences-in-depth'. As a result, the protection, detection and warning functions of these procedural safeguards

---

1   Eagle, C.J., Davies, J.M. and Reason, J. (1992). 'Accident Analyses of Large-Scale Technological Disasters Applied to an Anaesthetic Complication'. *Canadian Journal of Anaesthesia*, 39, pp. 118–22.

were invalidated, and the recovery and containment functions, though promptly applied, proved ineffective.

The Event

A 72-year-old man was listed to have a cystoscopy performed under local anaesthetic. Because the patient was booked for a local anaesthetic, an anaesthetist did not assess him beforehand.

On the morning in question, the attending urologist discovered that he was double-booked in the operating theatre, so he asked a urologist colleague to carry out the procedure in his stead. This colleague, who had not seen the patient, agreed, but insisted that the procedure should be carried out under a general anaesthetic.

The second urologist contacted the anaesthetic coordinator who found an available anaesthetist. But this anaesthetist was not told of the original booking under local anaesthesia or that the urologist performing the procedure was unfamiliar with the patient. The anaesthetist merely understood that the patient was an urgent addition to the operating theatre list.

The anaesthetist first saw the patient in the operating theatre, where he was found to be belligerent, confused and unable to give a coherent history. The nursing notes showed that the patient had fasted for 24 hours, but his chart also revealed several complications. These included progressive mental deterioration, metastatic cancer in the lungs and liver, renal insufficiency and anaemia.

Since the patient refused to be moved to the cystoscopy table, the anaesthetist decided to induce anaesthesia with the patient in his bed. Routine monitors were attached and Thiopentone (150 mg) and alfentanil (500 µg) were administered intravenously. The patient lost consciousness, but then regurgitated more than two litres of fluid and undigested food. He was immediately turned on his side and the vomitus was sucked from his mouth. The patient's lungs were ventilated with 100 per cent oxygen and intravenous fluids were given. Investigation showed large quantities of fluid in the patient's bronchi. He was transferred to the Intensive Care Unit and died six days later.

Analysis

Unsafe Acts

Active failures were committed by two people. These errors are listed below in temporal order:

1. The first urologist committed a rule-based mistake when booking the patient for cystoscopy under local anaesthetic. By applying the rule of thumb that 'elderly patients are given a local anaesthetic for cystoscopy', the attending urologist overlooked the counter-indications, namely that this was a confused and combative patient, and selected the wrong form of anaesthetic.

2. The anaesthetist similarly made a series of rule-based mistakes in assuming that the fasting status indicated on the nursing notes was correct, in assuming that all the relevant information was available on the chart and in deciding to proceed with the anaesthetic, despite having performed a fairly cursory pre-operative assessment on a patient with several medical problems.

Error-Enforcing Conditions

The anaesthetist, in particular, was the unwitting focus of a number of resident pathogens in the system:

1. Failures in the operating theatre scheduling process caused the first urologist to be double-booked.

2. Breakdowns in communication caused the anaesthetist to be unaware of the substitution of the second urologist and of the original booking for the procedure to be performed under a local anaesthetic.

3. This opened a serious breach in the pre-operative defences and resulted in the patient not being assessed by an anaesthetist on the evening before surgery.

4. In addition, a post-event review of the patient's chart revealed that an episode of projectile vomiting had occurred at 4.00 am on the morning of the procedure. This information

was recorded by a nurse in the computerized record-keeping system, but had not been printed out and attached to the chart due to lags in the system.

5. Furthermore, no terminal was available in the operating theatre, so the anaesthetist could not, in any case, have checked the computerized record for recent updates.

6. The anaesthetist was unaware of the delay in producing hard copies of the patient's record and, most importantly, was thus ignorant of the vomiting incident which would have created strong suspicion as to the 24-hour fasting claim.

Organizational Failures

Several familiar organizational failure types[2] featured prominently in the factors contributing to this adverse event:

1. Most obviously, there were significant *communication failures* at a number of points in the system.

2. There were deficiencies in the operating theatre *management system* that allowed the first urologist to be double-booked.

3. The *design and operation* of the computerized record-keeping system allowed crucial information to be denied to those at the 'sharp end'.

4. It is also likely that the absence of a terminal in the operating theatre was the result of a *goal conflict* between cost-cutting and patient safety.

Conclusion

This event illustrates very clearly how a number of active failures and latent conditions can combine to create a major gap in the system's defences. In such a multiply defended system, no one of these failures would have been sufficient to cause the patient's death; each provided a necessary link in the causal chain.

The railway accident, discussed next, further illustrates the leading role played by communication failures in the aetiology of an organizational accident.

---

2   Reason, J. (1990). *Human Error*. Cambridge: Cambridge University Press.

Case Study 2: The Glenbrook Rail Accident[3]

The Event

At 8.22 am on 2 December 1999, an inter-urban commuter train (designated as W354) crashed into the rear of an Indian Pacific (IP) cross-continental tourist train. The accident occurred on the main western line east of Glenbrook railway station in the Blue Mountains west of Sydney. The IP was on the last leg of its journey across Australia from Perth to Sydney, while the inter-urban train, which originated in Lithgow, was taking morning commuters into Sydney.

Below are listed the stations of relevance along the line going towards Sydney (main stations are in upper case):

- LITHGOW.
- KATOOMBA.
- SPRINGWOOD.
- Valley Heights.
- Warrimoo.
- Blaxland.
- Glenbrook (where the accident happened).
- Lapstone.
- EMU PLAINS.
- PENRITH (where the signalman was located).

A fault had occurred during the night in the automatic signalling system causing two signals – 41.6 (just to the east of Glenbrook) and 40.8 (the next signal along the track towards Sydney) – to failsafe; that is, they went to the red or stop aspect. Since the signals were not functioning normally, the Penrith signaller and the train drivers managed control of train movements in the area.

Both the drivers of the IP and W354 contacted the signalman on stopping at the red aspect signal. The signalman gave authority for both trains to pass 41.6 – at that time it was not known that 40.8

---

3 The Honourable Peter Aloysius McInerney. *Interim Report of the Special Commission of Inquiry into the Glenbrook Rail Accident.* Sydney: Special Commission of Inquiry, June 2000.

had also failed in the stop aspect. The IP (the first train) stopped at 40.8 in response to the red signal. When the signalman allowed the IU to pass 41.6, both the signalman and the driver believed that the track ahead was clear of the IP train. However, for reasons that will become clearer later, the IP was halted around a bend and out of sight of the oncoming W354. Train W354 crashed into a 'motorail' car at the end of the IP and seven passengers were killed instantly. The driver of W354 managed to avoid serious injury as he ran back through the first carriage when a collision appeared imminent. Dozens of passengers on both trains sustained injuries – 50 of them were taken to hospital by ambulance.

The Analysis

The Defences that Failed

The two main defensive failures were the automatic signalling system and the subsequent communications between the drivers and the signalman. There were gaps in both sets of defences, but none of them were very large. As in the first case study, the accident slipped through the small cracks in the system while everyone was trying to do their best.

Once the diode had failed (possibly due to a lightning strike), causing both signals – 41.6 and 40.8 – to fail in the red position, the passage of IP and W354 became controlled by a number of individuals. Both drivers, two on the IP and one on W354, were experienced individuals with nothing in their records to cast any doubt on their competence. The two other people directly involved were the Penrith signalman and his immediate supervisor, the Senior Controller at West Control (WCS). While the former was responsible for the signals in his immediate area, the latter organized and controlled trains through a larger area.

When the IP arrived at Glenbrook station at 8.04 am, both drivers expected 41.6 to be at red because the previous signal (42.0) had been at caution. The No. 2 driver put on his high-visibility jacket and walked across to the signal telephone adjacent to 41.6. But, unusually, he found the box locked and had to go back to the cab for his key. This caused a delay of some minutes. With the box now unlocked, he picked up the handset and heard someone say 'Penrith here'. The second driver told the Penrith signalman that 41.6 was in the stop position. The signalman replied: 'You're okay to go past the signal at stop and can you please report to me the indication of the next signal that you come across'.

When the IP driver got back into the cab, he said to his colleague (driving the train): 'I've got hold of the signalman at Penrith and he has okayed us to go past this signal. We will proceed ahead'.

They then proceeded to the next signal with extreme caution and it took nearly eight minutes to travel from signal 41.6 to signal 40.8. By this time, the train was some 15 minutes behind schedule.

At 8.09 am, the Penrith signalman (PS) called his supervisor at West Control (WCS). The interchange went as follows:

> WCS: Penrith.
> PS: *Oh, yeah, what have we got for you now, buddy? 41.6 signal at Glenbrook reported as a failure by the Indian Pacific. He has been delayed there for five minutes.*
> WCS: It's only an auto, isn't it?
> PS: *It is only an auto, mate, that's correct, but the driver decided to call me five minutes later.*
> WCS: Let me just find it here, 41.6.
> PS: *Yeah, mate. I will let Springwood know if you could get hold of an electrician for us. We've got a bloody stack of dramas down here this morning.*
> WCS: Yeah, okay … what time did he tell you that?
> PS: *They told me at 10 minutes past, mate.*
> WCS: Any other signal out? Just that one was it?
> PS: **As far as I know, mate. I will ask him when he gets to Emu Plains if that was the only signal, mate.**

PS's last remark (in bold) is important because it confirms that he was unaware that the next signal, 40.8, had also failed. The fact that he had not yet had any report from the IP that 40.8 was failed may have led him to believe that there was nothing wrong with 40.8.

Shortly after this conversation, W534, the inter-urban, was approaching Blaxland, the station before Glenbrook. This train was fitted with MetroNet radio system, so that the driver of W534 could communicate directly with PS without leaving his cab. Ironically, the IP also had a two-way radio (but not MetroNet);

however, the Safe Working Instruction 245 required that all trains without MetroNet should communicate by signal telephone (on the side of the track). Had the IP driver used his radio rather than the telephone, a valuable five minutes between the IP and the inter-urban could have been saved. As will be seen later, this factor is even more critical when the IP stops at 40.8.

At 8:12, the West Control supervisor (WCS) radioed the cab of W534 and the following conversation ensued:

> WCS: West Control to the driver of W534.
> W534:   *Yes, mate.*
> WCS: I've just had a report from the driver of the Indian Pacific ahead of you, 41.6 signal, Penrith side of Glenbrook platform is at stop.
> W534:   *Oh, right.*
> WCS: Are you around that area yet?
> W534:   *I'm just heading towards Blaxland now, mate.*
> WCS: Heading towards Blaxland, all right, okay. Well, I don't know what – he just said it's failed, so I don't know if it's a red marker light or what the story is. It's only an auto.
> W534:   *All right mate. I'll get onto Penrith anyhow or whatever, whoever I get there.*
> WCS: Yeah, okay.
> W534:   *Thanks a lot.*
> WCS: **Well, yeah, it's only an auto so just trip past it.**
> W534:   *Okay, thanks.*

The phrase 'just trip past it' was ill-advised, giving the W534 driver an expectation that the track ahead would be clear. It was evident that the driver knew that 41.6 had failed. He also believed that all of the preceding trains would be clear of the immediate area once he had passed that signal. His mind was conditioned to the belief that there were no trains between signal 41.6 and 40.8 – this was apparent from the manner in which he spoke to the Penrith signalman when he radioed him at 8.20 am:

> PS: *Yeah, 534?*
> W534:   Yeah, who have I got there, mate?
> PS: *Penrith, mate.*

W534:   Yeah, it is 41, 41.6, **I'm right to go past it, am I, mate?**

PS: ***Yeah, mate, you certainly are.*** *Listen, can you get back to us? What was the previous signal showing?*

W534:   Yellow.

PS: *Yellow, okay, and what's that signal exactly showing, just red or ...?*

W534:   **Yeah, two reds, mate.**

PS: ***Two reds, no worries. All right mate, can you let us know what the signal in advance says when you get to it? Thanks.***

W534:   Okay, matey.

PS: *Okay, thanks.*

This amiable piece of dialogue contained some traps for the unsuspecting driver of W534. It reinforced his belief that the line ahead was clear – especially those remarks in bold. They were effectively a verbal green light from the signalman.

Meanwhile, the IP, far from being cleared from the track ahead, was halted at the failed red signal 40.8. Once again, the second driver left the cab to contact the Penrith signalman on the signal telephone – as instructed by the Safety Work Order. When he got there, he found the 'push to ring' button missing. He attempted to contact the Penrith signal box by turning the crank handle five or six times, but could get no coherent reply, though he did hear voices at the other end. The reasons for this malfunction are still not established. The telephone was found to be working on the afternoon of the accident.

W534, having passed 41.6, accelerated to a speed of 40–50 kph. The driver said that this was well under what would have been the normal speed for that section of track. The driver's sighting distance of the rear of the IP was probably around 135 metres, and its speed at the time of the collision was between 30–40 kph.

Additional System Problems

The signaller at the Penrith signal box lacked a train indicator board and thus had no means of establishing the position of a train in his area other than by communications with the drivers or other railway employees.

The insistence of the Safe Working Instructions upon using signal telephones, even when the IP was equipped with a working two-way radio, was a major contributor to this accident, as were the problems encountered by the second IP driver when trying to use the signal telephones at both 41.6 and 40.8.

Conclusion

This case study emphasizes the importance of unambiguous and standardized communications in safety-critical operations – something the aviation industry learned a long time ago. Everyone in this story was trying to be helpful; everyone was trying to do a good job. But between all of these good intentions, some fatal ambiguities crept in. It is easy to see this after the event. At the same time, it is hard to imagine a set of amiable, competent, experienced Australian railwaymen speaking to each other any differently during what appeared to them to be a routine working day. Unfortunately, the system in this case was not at all forgiving of these minor lapses.

Case Study 3: The Omnitron 2000 Accident[4]

This event differs from the usual run of medical mishaps in that it had adverse consequences for over 90 people. The accident, unusually, was also reported in the public domain by the US Nuclear Regulatory Commission. It occurred as the result of a combination of procedural violations (resulting in breached or ignored defences) and latent failures.

The Event

In 1992, an elderly patient with anal carcinoma was treated with high dose rate (HDR) brachytherapy. Five catheters were

4   NUREG (1993). *Loss of an Iridium-92 Source and Therapy Misadministration at Indiana Regional Cancer Center, Indiana, Pennsylvania, on November 16th 1992.* Report 1480. Washington, DC: US Nuclear Regulatory Commission. Vincent, C. and Reason, J. (1999). 'Human Factors Approaches in Medicine'. In: M. Rosenthal, L. Mulcahy and S. Lloyd-Bostock (eds), *Medical Mishaps: Pieces of the Puzzle.* Buckingham: Open University Press.

placed in the tumour. An iridium-192 source (4.3 curies or 160 gigabecquerels) was intended to be located in various positions within each catheter, using a remotely controlled Omnitron 2000 afterloader. The treatment was the first of three treatments planned by the physician, and the catheters were to remain in the patient for the subsequent treatments.

The iridium source wire was placed in four of the catheters without apparent difficulty, but after several unsuccessful attempts to insert the source wire in the fifth catheter, the treatment was terminated. In fact, a wire had broken, leaving an iridium source inside one of the first four catheters. Four days later, the catheter containing the source came loose and eventually fell out of the patient.

It was picked up and placed in a storage room by a member of the nursing home staff, who did not realize it was radioactive. Five days later, a truck picked up the waste bag containing the source as part of the driver's normal routine. It was then driven to the depot and remained there for a day (Thanksgiving) before being delivered to a medical waste incinerator, where the source was detected by fixed radiation monitors at the site. It was left over the weekend, but was then traced to the nursing home. It was retrieved nearly three weeks after the original treatment. The patient died five days after the treatment session and, in the ensuing weeks, over 90 people were irradiated in varying degrees by the iridium source.

Analysis

Active Failures

1. The area radiation monitor alarmed several times during the treatment, but was ignored – partly because the physician and technicians knew that it had a history of false alarms (cf. the obstetric incident presented earlier).
2. The console indicator showed 'safe' and the attending staff mistakenly believed the source to be fully retracted into the lead shield.
3. The truck driver deviated from company procedures when he failed to check the nursing home waste with his personal radiation survey meter.

Latent Conditions

1. The rapid expansion of HDR brachytherapy from one to 10 facilities in less than a year had created serious weaknesses in the radiation safety programme.
2. Too much reliance was placed upon unwritten or informal procedures and working practices.
3. There were serious inadequacies in the design and testing of the equipment.
4. There was poor organizational safety culture. The technicians routinely ignored alarms and did not survey patients, the afterloader or the treatment room following HDR procedures.
5. There was weak regulatory oversight. The Nuclear Regulatory Commission did not adequately address the problems and dangers associated with HDR procedures.

Conclusion

This case study illustrates how a combination of active failures and latent systemic weaknesses can conspire to penetrate the multi-layered defences designed to protect both patients and staff. No one person was responsible for the disaster. Each individual acted according to his or her appraisal of the situation. Yet one person died and over 90 people were irradiated.

Case Study 4: Loss of Engine Oil on a Boeing 737-400[5]

On 23 February 1995, shortly after departing from East Midlands Airport en route for Lanzarote in the Canary Islands, the pilots of a Boeing 737-400 detected the loss of oil quantity and oil pressure on both engines. They declared a 'Mayday' and diverted to Luton Airport, where both engines were shut down during the landing roll. There were no casualties. It was later discovered that the high-pressure rotor drive covers on both engines were missing, resulting in the almost total loss of the oil from both engines

---

5   Air Accident Investigation Branch (1996). *Report on the Incident to Boeing 737-400, G-OBMM, Near Daventry on 23 February 1995.* AAIB Report 3/96. London: HMSO.

during flight. A scheduled borescope inspection (required every 750 hours) had been carried out on both engines during the previous night by the airline's maintenance engineers.

Background

The borescope inspections were to be performed by the line maintenance night shift. On the night before this work was to be done, the Line Engineer in charge of this shift had expressed his concern about the manpower available to carry out the next night's predicted workload, which he knew would include the borescope inspections. However, on arriving for work on the night in question, he discovered that no extra maintenance personnel had been assigned to his shift: instead of a nominal complement of six, there were only four on duty that night and two of them – including the Line Engineer – were doing extra nights to cover shortfalls.

The Line Engineer realized that he would have to carry out the borescope inspections himself. He was the only engineer on the line shift possessing the necessary authorization. Since the inspection was to be carried out in a hangar at some distance from where the bulk of the line maintenance work was done, but close to the base maintenance hangar, he decided to put the inspection at the top of his list of jobs for that night.

After organizing the night's work for his shift, he collected the inspection paperwork from the line office and went to the aircraft, where he started to prepare one of the engines for inspection. Having done this, he went across to the nearby base maintenance hangar where the borescope equipment was stored. There he met the Base Maintenance Controller (in charge of the base night shift) and asked for the inspection equipment and also for someone to help him, since the engine spool had to be turned by a second person as he carried out the inspection.

The base night shift was also short-staffed. On arriving at his office that night, the Base Controller had received a request from Ramp Services to remove another aircraft from their area (a B 737-500) because it was in the way. He could not do this immediately because of the shortage of personnel on his shift. However, when he met the Line Engineer, he saw a way of killing two birds with one stone. The company rules required staff to carry out two 750-hour borescope inspections within a 12-month period in order to maintain their borescope authorization. His own authorization was in danger of lapsing because this task very rarely came his way. He also knew that the line maintenance shift was lacking two people and that they had eight aircraft to deal with that night. So he offered to do a swap. He would carry out the borescope inspection if the Line Engineer took over the job of moving the 737-500 from the Ramp to Base.

The Line Engineer agreed to this mutually convenient arrangement and told the Base Controller where he had got to in the preparation of the engines. Since the Line Engineer had not gone to the Base Hangar with any intention of handing over the inspection to someone else, there was no written statement or note of what had been done so far. Indeed, the line maintenance paperwork offered no suitable place to record these details. The Base Controller, however, was satisfied with the verbal briefing that he had received. His next step was to select a fitter to assist him and to check and prepare the borescope equipment.

With these innocent and well-intentioned preliminaries, the stage was set for a near-disaster. The fitter was sent to prepare the second engine for inspection and the Base Controller then spent a considerable time organizing the duties for his short-staffed night shift. When he eventually joined the fitter at the aircraft, he brought with him his personal copy of the borescope training manual on which he had written various details to help him in judging sizes through the probe. Although the work pack for the job was unfamiliar to him – being Line rather than Base maintenance paperwork – and although there are no Boeing task cards attached, he did not consider it necessary to draw any additional reference material. While the fitter was continuing to prepare the second engine for inspection, the Base Controller started his inspection of the first engine.

Throughout the next few hours, the job continued with many interruptions brought on by the Base Controller's need to cope with pressing problems cropping up elsewhere on his patch. The upshot was that the rotor drive covers were not refitted, the ground idle engine runs (checks that could have revealed the oil leaks) were not carried out and the borescope inspections on both engines were signed off as being complete in the Aircraft Technical Log.

During the night, the Line Engineer returned with a line maintenance crew to prepare the aircraft for the next day's flying. He and the Base Controller discussed whether or not to trip out the aircraft's engine ignition and hydraulic circuit breakers. The Base Controller said he did not think it was necessary since he had no intention of working on the aircraft with the electrical and hydraulic power systems active, but the Line Engineer pulled the circuit breakers anyway, feeling confident that the Base Controller was aware of this. When the aircraft was later returned to the line, the pilots noted that these circuit breakers had not been reset and expressed some dissatisfaction about this to a line engineer on the morning shift. The engineer said that it was probably an oversight and reset them. When the engineer returned to the line office, he wondered aloud to a colleague how the engine run had been done with the ignition circuits disabled. They were still discussing this when they heard that the aircraft had made an emergency landing at Luton.

We do not know exactly why the covers were not replaced on the engines, or how the inspection came to be signed off as completed when such vital parts were missing and the engine run was not done. We do know, however, that omissions during re-assembly are the single most common form of maintenance lapse – both in aviation and in the nuclear power industry. We know too that there had been at least nine previous instances in which high-power rotor covers had been left off engines following maintenance in other airlines. The available evidence also indicates that many aircraft maintenance jobs are signed off as complete without a thorough inspection. Such routine shortcuts are most likely to occur in the face of a pressing deadline – like getting an aircraft back onto the line before its scheduled departure time.

Analysis

What, then, were the small cracks in the system that permitted these commonplace deviations to come so close to causing a major air disaster? The first and most obvious factor is that both the line and base maintenance crews were short-staffed, though not so depleted that they had to abandon their assigned tasks. Such temporary shortfalls in maintenance personnel through sickness and leave are not uncommon. A similar problem contributed to the blow-out of a flight deck windscreen that had been secured by the wrong-sized bolts. There, as in this case, a temporary shortage of personnel led supervisory staff to take on jobs for which they lacked the current skills and experience. And, in both events, their ability to perform these tasks reliably was impaired by the continuing need to carry out their managerial duties. To compound these problems further, the shift managers involved in both accidents were also in a position to sign off their own incomplete work, thus removing an important opportunity for detecting the error.

In the case of the missing engine covers, however, there was also another more subtle crack in the system – the difference in working practices between line and base maintenance. The division of labour between these two arms of an aircraft maintenance facility is normally clear-cut. Line engineers are accustomed to performing isolated and often unscheduled tasks (defect repairs) either individually or in small groups. These jobs are usually completed within a single shift. Base maintenance, on the other hand, deals with scheduled overhauls. When an aircraft enters a base maintenance hangar, it is likely to be there for a relatively long time and a large number of different tasks will be carried out by rotating shifts, with incomplete work being handed on from one shift to the next.

The line and base environments thus require different kinds of planning and supporting work packs. In the case of line maintenance, the paperwork is generated just before the work is due to be done and is subject to change according to operational needs. The job cards give only a brief description of the work to be performed. In sharp contrast, planning for base maintenance work starts several weeks before the event and is delivered as a massive work pack that includes a large amount of supporting and explanatory documentation. Life on the line is full of surprises, while that in base maintenance is far more regular and predictable.

The 750-hour borescope inspection was designated as a line maintenance task, and the paperwork included none of the step-by-step task cards and maintenance manual sections normally supplied with a base maintenance work pack; it only contained references to the detailed documentation that was available elsewhere. The Line Engineer was familiar with the job and did not feel it necessary to draw this additional information, so the work pack that he handed over to the Base Controller was relatively sparse compared to the usual base maintenance documentation. In particular, the line-generated work pack contained no mention of the restorative work – replacing fasteners and covers – nor did it require step-by-step signatures confirming the completion of these tasks, as would be expected in a base-generated work pack. Nor were there any of the customary warnings – highlighted in the maintenance manual – that the safety of the aircraft would be seriously jeopardized if the re-assembly work was not completed as specified.

Given the short-staffed nature of his own shift and the additional demands that this would inevitably make upon his time, the Base Controller was probably unwise to offer to carry out the borescope inspection. But it could hardly be classed as a major blunder and it was certainly allowable within company procedures. In view of his lack of current experience at the task, he was clearly mistaken in assuming that he could perform the task without reference to detailed task cards and using only his memory and the unofficial training notes as a guide. Until he actually received the work package from the Line Engineer, he probably would not have known that the detailed supporting material was missing. At this point, he could have gone across to the line maintenance office or to base maintenance to get it, but it was a cold winter's night and he was confident that he knew the procedure. And even the accident investigator acknowledged that had he retrieved this material, it would not have been easy to use. All manuals are continually being amended and updated, often making it hard for those unfamiliar with their layout to follow a continuous thread of task-descriptive text.

Another small crack in the system was the absence of any formal procedure for handing over jobs from line to base maintenance. As a result, no pre-prepared stage sheets (indicating the prior work performed) were available and the handover briefing was done verbally rather than in writing. This was not so much a case of a flawed defence as one that was absent entirely. Yet such handovers are comparatively rare events, and while it is easy to appreciate the need for handover procedures in retrospect, it is hard to imagine why their creation should be high on anyone's agenda before this event occurred. These various cracks in the defences are summarized in Table 6.1.

In July 1996, the airline involved in this accident was fined £150,000 plus £25,000 costs for 'recklessly or negligently'

Table 6.1   Summarizing the active failures and latent conditions that undermined or breached the aircraft maintenance system's defences

Active failures	Latent conditions
The line maintenance engineer who released the aircraft to the flight crew missed an opportunity to discover the missing drive covers when he reset the ignition circuit breakers and queried how an engine run could have been performed.	The airline lacked an effective way of monitoring and adjusting the available manpower to the workload, particularly on night shifts known to be subject to adverse time-of-day effects.
The flight crew accepted the pulled circuit breakers on the flight deck as 'normal' and did not pursue the matter further.	The regulatory surveillance of this manpower problem was inadequate.
The aircraft's Technical Log was signed off without adequate inspection of the work on the engines.	Both the line and the maintenance night shifts were short-staffed. This was a fairly common occurrence.
The Base Controller was in a position to sign off his own incomplete work without further inspection.	An internal inquiry revealed that bore-scope inspections were often carried out in a non-procedural manner. This was a failure of the Company's Quality Assurance system.
The Base Controller failed to keep a close enough eye on the work of the relatively inexperienced fitter. This was partly due to the many interruptions arising from his administrative duties.	No procedures existed for the transfer of part-completed work from line to base maintenance. As a result, no stage paperwork was available, and the Line Engineer only gave a verbal handover to the Base Controller.
No idle engine run was performed.	The line-oriented paperwork given to the Base Controller lacked the usual (for him) reference documentation. In particular, it lacked any reference to restorative work and provided no means of signing such work off.
The Base Controller carried out the work without reference to task cards or the maintenance manual. Instead, he relied on his memory and an unapproved reference – his training notes.	

endangering an aircraft and its passengers. This was the first time in the UK that the Civil Aviation Authority had prosecuted a carrier under Articles 50 and 51 of the 1989 Air Navigation Order. The two maintenance engineers directly involved in the accident were dismissed. After the verdict, the Deputy Chairman of the airline told the press: 'This has been a difficult day for us, but we have learnt from the experience. We have completely changed our procedures'.

The captain of the aircraft, whose prompt actions averted a catastrophe, said that it was a pilot's job to cope with the unexpected: 'We have to anticipate the worst case scenario. We are not just up there to press a button and trust in the wonders of modern technology. We have to be ready for this kind of eventuality'. This states the defensive role of flight crew very clearly.

Case Study 5: JCO Criticality Incident at Tokai[6]

The Event

On 30 September 1999, Japan suffered its worst criticality accident in the post-war period – an event that aroused an enormous amount of public anxiety in a country with good cause to fear the effects of radioactivity. It happened in the JCO uranium processing plant when three workers in the little-used conversion building poured an enriched uranium solution dissolved in nitric acid into a precipitation tank. The total amount poured reached 16.8 kg uranium by mass. The radiation monitors signalled a criticality event. Two of the three workers died: one, who had received a 17 Sv exposure, died 83 days later, while the other, who had received a 7 Sv exposure, died 211 days later.

The JCO plant had three processing facilities. There were two buildings for producing low-enriched uranium (less than five per cent) in which the bulk of their work was carried out. They had an annual capacity of 220 and 495 tons respectively. The building

6   Furura, K., Sasou, K., Kubota, R., Ujita, H., Shuto, Y. and Yagi, E. (2000). 'Human Factor Analysis of JCO Criticality Accident'. *Cognition, Technology & Work*, 2, pp. 182–203. See also comments on the JCO accident report in the same issue of the journal.

in which the accident occurred – the conversion building – was isolated from the rest of the facility. It had originally been designed for producing uranium powder of 12 per cent enrichment. However, its annual capacity was now only three tons. All the equipment within this facility, with the important exception of the precipitation tank (P-tank), had criticality-safe geometry – that is, it was shaped in such a fashion so as to prevent a criticality event. The P-tank was protected by mass control – that is, there was a strict limit on the amount of uranium that could be placed inside the tank at any one time. The only customer for this facility was the JOYO experimental fast breeder reactor. Orders were few and far between, so the facility was seldom used.

The Background

The accident arose from the gradual modification of working procedures over several years. The modifications of relevance to the accident concerned the homogenization process carried out in the conversion building. In the 1980s, the plant had been licensed by the regulatory authority to carry out the homogenization process using the following safe but slow procedure:

- One batch of purified uranium was dissolved in nitric acid in the dissolution tank. This solution was then transferred to five specially constructed five-litre bottles.
- This process was then repeated some six to seven times, creating 10 bottles of the uranium solution.
- The solution was homogenized using the cross-blending method in which a tenth part of each solution was taken from each bottle and placed in another 10 bottles.

In 1993, someone – in the spirit of *kaizen* – hit upon a way of speeding up this process. In the previous technique, the dissolution tank had been common to both the purification and homogenization processes. This bottleneck could be overcome if, instead of dissolving the uranium in the dissolution tank, they used large stainless steel buckets to achieve this end.

In 1994–6, it was decided to extend this innovation by using the bucket method for dissolving the uranium in the purification process as well. This shortened the production time by more than 50 per cent. The purified uranium was dissolved using the buckets, and then they decided to one of the two buffer columns to homogenize the uranium solution. To achieve this, they rigged a makeshift line to isolate the buffer column and pump. They could then put six to seven batches into the buffer column at one go because it had safety-critical geometry.

Although this modification greatly increased production time, it had a snag: the outlet from the buffer column was inconveniently located, being very close to the floor. On the day before the accident, the team performing the task had a further idea about how to increase production – a fatal one, as it turned out. Instead of using the buffer column with its inconvenient access, they decided to use the much larger P-tank. The plan was as follows. They would dissolve the purified uranium in nitric acid using a stainless steel bucket. This would be filtered and transferred successively to a five-litre beaker, which would then be poured through a funnel into an observation hole at the top of the P-tank (another improvization). The P-tank was not only larger, it had a mechanical stirrer to speed the homogenization along – what it did not have, however, was criticality-safe geometry. Thus, the stage was set for a disaster towards which the whole system had been unknowingly inching its way for over a decade.

The Immediate Causes

We can summarize the immediate causes in a sentence. The wrong workers following the wrong advice and using the wrong methods poured the wrong concentration of uranium into the wrong vessel at the wrong time. It took many years to combine so many 'wrongs' in the same place at the same time. Let's look at each 'wrong' in detail.

The Wrong Workers

The JCO company was in economic difficulties. A downsizing programme had begun in 1996 and many workers, both administrative and operational, had been laid off. There had been close to a 50 per cent reduction in the production division, with the workforce reduced from 68 to 38 – and a lot of know-how had departed with them. None of the three workers assigned to carry out the job that resulted in the criticality accident had had any experience in the conversion building. As indicated earlier, it was rarely used.

To exacerbate the problem further, JCO had ceased giving criticality training to its production workforce because it had complained that it was too difficult to understand. The company relied instead upon step-by-step procedures. But these procedures were out of date and carried no instructions about safety. In any case, the workforce rarely used them. There was also very little in the way of informed supervision, particularly in the conversion building.

The Wrong Advice

The team leader of the ad hoc group assigned to carry out the task in the conversion building had, on the day prior to the accident, seen a member of the planning group at lunch in the canteen. He had asked him whether it would be all right for them to use the P-tank for the homogenization. He said it was okay.

Later, he said he had confused the conversion room with the more widely used fabrication plants. He also said he had forgotten that high-enriched uranium rather than low-enriched uranium was to be used.

The Wrong Method

The government licence required the homogenization process in the conversion building to be done in the dissolution tank. As noted earlier, however, it had been JCO's practice since 1993 to dissolve purified uranium in stainless steel buckets. This cut production time because the workers could put the uranium solution in the tank of their choice for the homogenization process. In two previous orders, they had used an isolated buffer column with criticality-safe geometry, allowing the processing of six to seven batches. The managers realized that this failed to comply with their licence, but, nonetheless, they had issued modified procedures that described this illegal method.

The Wrong Concentration

At the time of the accident, the workers had poured 16 kg of uranium (19 per cent enrichment). Four batches were poured during the afternoon preceding the accident. The mass was then 9.6 kg. The next morning, they poured three further batches into the P-tank. The criticality accident occurred at 10.35 am while the seventh batch was being poured.

The Wrong Vessel

As noted earlier, the workers on this occasion elected to use the P-tank rather than the buffer column because the latter had inconvenient access. But the P-tank did not have criticality-safe geometry; it required mass control which they did not understand.

The Wrong Time

The workers deviated from the work plan and sped up the purification process. Purification for all 54 kg was completed eight days ahead of schedule. The stir-propeller in the P-tank was seen as a means of hastening homogenization. They estimated that it would take two rather than four days to complete the process using the P-tank. They were hurrying because they wanted the order completed before on-the-job training in waste processing started in October.

The Failed Defences

There were five distinct failures of the system's protection:

- *Training and education*: comprehension problems on the part of the workforce led to the abandonment of criticality education several years earlier.
- *Procedures*: the procedures focused almost exclusively upon production issues and failed to provide any safety-related information.
- *Batch control*: this was 'forgotten' several years earlier due to production demands.

- *Mass control*: this was lost with the buffer column method of homogenization.
- *Criticality-safe geometry*: this was bypassed when the workers elected to use the P-tank rather than the buffer column.

The whole dismal story can be interpreted as the successive losses of protection in favour of increased rates of production. But, in each case, the step was an isolated one. It was only on 30 September 1999 that all of these insidiously acquired 'fault-lines' in the defences came together to produce the criticality accident.

Conclusions

With the benefit of hindsight, it is possible to see this progress to disaster as inexorable in view of the almost complete lack of a safety culture at all levels of the JCO company. Given the dwindling profits, the intermittency of the orders for work in the conversion room, the lack of adequate supervisory oversight, the successive illegal modifications, the erosion of respect for the very considerable hazards and the lack of safety training and procedures, it was probably inevitable that some work group would see the production benefits of using the P-tank and thus breach the final mass control defence.

Case Study 6: A Vincristine Tragedy

The Event

A close examination of this adverse event, which occurred on 4 January 2001, is possible because its organizational precursors were investigated by an external expert in accident causation and the very detailed findings were made available to the public domain (Toft, 2001).[7]

---

7   Toft B. (2001). *External Inquiry into the Adverse Incident that Occurred at Queen's Medical Centre, Nottingham, 4th January 2001*. London: Department of Health; Department of Health. (2000). *An Organization with a Memory. Report of an Expert Group on Learning from Adverse Events in the NHS Chaired by the Chief Medical Officer*. London: Stationery Office, p. 25.

An 18-year-old male patient, largely recovered from acute lymphoblastic leukaemia, mistakenly received an intrathecal injection of the cytotoxic drug vincristine. The treatment was given by a senior house officer (SHO) who was supervised by a specialist registrar (SpR). The former was unfamiliar with the usually irreversible neurological damage caused by the intrathecal administration of vincristine, and the latter had only been in post for three days. It was a requirement that the spinal administration of drugs by SHOs should be supervised by an SpR. This supervisory task fell outside the scope of the SpR's duties at that time, but no one else seemed to be available and he wanted to be helpful. The error was discovered very soon after the treatment and remedial efforts were begun almost immediately, but the patient died just over three weeks later.

The hazards of injecting vincristine intrathecally (rather than intravenously) were well known within the prestigious teaching hospital where this tragedy happened. This particular adverse event has occurred several times before. An influential report commissioned by the UK's Chief Medical Officer featured such an accident as a full-page case study.[8] It was noted that there had been 14 similar events in the UK since 1985. Other surveys indicate that a large number of such occurrences have occurred worldwide.

The precise numbers are not important here; what matters is that the same procedure has been directly associated with iatrogenic fatalities in a large number of healthcare institutions in a variety of countries. The fact that these adverse events have involved different healthcare professionals performing the same procedure clearly indicates that the administration of vincristine is a powerful error trap. When the same situation repeatedly provokes the same kind of error in different people, it is clear that we are dealing with an error-prone situation rather than error-prone, careless or incompetent individuals.

The hospital in question had a wide variety of controls, barriers and safeguards in place to prevent the intrathecal injection of vincristine. But these multiple defences failed in many ways

---

8   Donaldson, L. (2000). *An Organization with a Memory*. London: Department of Health.

and at many levels. The 'upstream' defensive breakdowns and absences are summarized below.

The Analysis

Administrative and Procedural Measures that Failed

- The hospital medical staff had modified the protocol on which the patient's treatment was based so that vincristine (intravenous route) and cytosine (intrathecal route) were to be administered on different days. (The original protocol allowed them to be given on the same day.) In an effort to reduce the inconvenience to patients, the nursing staff had adopted the practice of bringing both drugs to the ward on the same day.
- The amended protocol required the intrathecal injection of chemotherapy to be given on the first day and the intravenous vincristine on the second. On the patient's prescription, however, vincristine was entered as the first item, though it was to be administered on the second day.
- Within the pharmacy, it was required that there should be separate labelling, packaging and supply of intrathecal (IT) and intravenous (IV) cytotoxic chemotherapy to ensure that the drugs to be administered by different routes should not arrive at the day case unit at the same time and in the same package. On this occasion, however, the pharmacy allowed the drugs to be released together in the same clear bag.
- The hospital's *Drug Custody and Administrative Code of Practice* contained important information on good practice with regard to the prescription of drugs and the responsibilities of the medical staff in relation to the safe preparation, checking, administration and recording of drugs. But the supervising consultants were unaware of its existence and its safety-critical contents had not been brought to the attention of the SpR and the SHO.
- The *Haematology Conditions and Protocols* issued on the ward stated that IV and IT chemotherapy drugs should not be together on the same trolley, that methotrexate, cytosine and hydrocortisone were the only drugs to be administered

intrathecally (spinally), and that these drugs should never be given on the same day as IV vincristine. But two versions were on the ward at the same time, and this last instruction was missing from the one given to the SpR. In addition, these guidelines made no mention of the usually fatal consequences of giving vincristine intrathecally. There is no formal record of whether the SpR received a copy of the protocols and guidelines.

Indicators and Barriers that Failed

- The same prescription form was used for both intrathecal and intravenous cytotoxic chemotherapy. The only indication of their distinctive routes was given by handwritten initials rather than by separate forms that are clearly distinguished in colour.
- The IV route for the vincristine on the 'regular prescription' section of the chart was written in a barely legible fashion, at least in comparison to the other route instructions given on the same chart.
- The name of the drug and the dosage was printed in 9-point bold type on the syringe labels. The warnings with regard to route of administration were given in 7-point normal type, thus de-emphasizing this critical information.
- The warning printed on the side of the vincristine syringe – 'Vincristine Sulphate Injection 2mg/2ml NOT FOR INTRATHECAL USE' – was partially obscured by the attachment of the label, again de-emphasizing its significance.
- Although there were physical differences between the two syringes (i.e., different-coloured protective caps – grey for vincristine and red for cytosine), there were also a number of similarities: the syringes were of similar size; the respective volume of the two drugs (2 ml for vincristine and 2.5 ml for cytosine) offered little in the way of discriminatory cues; the contents of both syringes were colourless fluids; and, most significantly, both syringes could be connected to the spinal needle delivering the intrathecal drugs.

Failures of Supervision and Instruction

- Given the largely apprenticeship style of postgraduate medical training, there is a heavy responsibility upon senior medical staff to ensure that junior doctors and those newly appointed are clearly informed as to the nature and scope of their duties, as well as ensuring that they are aware of all the relevant protocols, guidelines and codes of practice. It is also essential that these supervisors and mentors should establish at the outset the extent of their charges' knowledge and experience with regard to potentially hazardous procedures. It is apparent that the induction and training of the new SpR and the SHO were far from ideal, and, most significantly, they were virtually non-existent with regard to the well-documented patient safety issues on the ward in question.
- The locum consultant asserted that he and the consultant haematologist had informed the SHO and other junior doctors present on a ward round two days earlier as to the dangers of the intrathecal administration of vincristine. This is contested by a number of the doctors present. Thus, there are grounds for doubting that this warning was ever given.

Communication Failures and Workarounds

- The locum consultant was concerned about the patient's history of poor time-keeping and treatment compliance, and had made a verbal request that he should be informed as soon as the patient arrived for his scheduled maintenance treatment on the day of his appointment. This request was not written down and was not acted upon.
- On the day after the SpR's arrival on the ward, the consultant haemotologist told him that he would have restricted clinical duties for the next two weeks and that he should 'shadow' a staff grade doctor in order to observe the workings of the ward. However, it soon became apparent that the precise or intended meaning of the term 'shadowing' was not understood by the SpR, the SHO or the nursing staff.

- The two drugs arrived on the ward in the same package. This occurred despite the fact that the pharmacist had made a note in the Sterile Production Unit log stating that the vincristine would be required on 'Thurs AM' and added the instruction 'send separately'.
- It had become common practice for ward staff to request the pharmacy to send both intrathecal and intravenous drugs together. The pharmacists complied because they did not want to be accused of compromising patient care.
- The day case coordinator on the ward left for home at 4.00 pm, not having informed the SpR that the patient and his grandmother had arrived. They were late. The treatment had been scheduled for the morning. Similarly, the ward sister went off duty at 4.30 pm without telling the SpR of the patient's arrival some 15–30 minutes earlier.

Collective Knowledge Failures and False Assumptions

- The staff nurse on the day case unit took a blood sample from the patient and then informed the SHO of his arrival. She also told the SHO that since an IT injection was to be given, he would need to be supervised by an SpR. Not appreciating the limited scope of the SpR's duties, the SHO approached the SpR and informed him that the patient was due to have an IT injection of chemotherapy.
- The SpR did not know that the SHO was unfamiliar with the patient's treatment and condition, and was ignorant of the dangers associated with the wrong-route administration of vincristine. He agreed to carry out the supervision, believing it to involve the simple provision of oversight for a junior colleague who knew the patient and understood the procedures.
- The two junior doctors asked the staff nurse where the patient's chemotherapy was located. Anxious to help, she eventually found it in the day case unit refrigerator. The transparent plastic package containing both syringes was the only item of chemotherapy in the refrigerator. She checked that the patient's name was printed on each of the

syringe labels and then took the package to the treatment room, where the SpR was alone. She handed the package to him with the words: 'Here's X's [the patient's first name] chemo'. Although a trained chemotherapy nurse, she did not herself check the nature of the drugs or their routes of administration. She assumed that the SpR and the SHO would do the necessary checking. She also assumed that both doctors were experienced in the administration of the chemotherapy drugs.

The Situation Just Prior to the Injections

At 5.00 pm, 20 minutes before the drugs were administered, the vast majority of the ingredients for the subsequent tragedy were in place. The many gaps and absences in the system's multiple 'upstream' defences had been unwittingly created and were lining up to permit the disaster-in-waiting to occur. Two inadequately prepared junior doctors, each with inflated assumptions about the other's knowledge and experience, were preparing to give the patient his chemotherapy.

It was a Thursday afternoon, which was normally a quiet time on the ward. The locum consultant was working in his office; the staff grade doctor whom the SpR was supposed to shadow was a part-timer and was not on duty that day. The ward sister had gone home. There were no other SpRs available that afternoon. There was no senior medical presence in the vicinity to thwart a sequence of events that was now very close to disaster. To compound the situation further, the patient and his grandmother had arrived unannounced and unscheduled for that particular time. The last 'holes' were about to move into alignment.

The Last Line of Defence: The Junior Doctors on the Spot

The SHO had wanted to administer the drugs in order to gain experience in giving spinal injections. The SpR handed him the syringes. In doing this, he read out the patient's name, the drug and the dose from the syringe label. He did not read out the route of administration. There were also other omissions and errors:

- He failed to check the treatment regimen and the prescription chart with sufficient attention to detect that vincristine was one of the drugs in the packet and that it should be delivered intravenously on the following day.
- He failed to detect the warning on the syringe.
- He failed to apprehend the significance of the SHO's query – 'vincristine intrathecally?' – on being handed the second syringe.

These errors had grievous consequences. But the SpR's actions were entirely consistent with his interpretation of a situation that had been thrust upon him, which he had unwisely accepted and for which he was professionally unprepared. His perception that he was required to supervise the intrathecal administration of chemotherapy was shaped by the many shortcomings in the system's defences. He might also have reasonably assumed that all of these many and varied safeguards could not have all failed in such a way that he would be handed a package containing both intravenous and intrathecal drugs. Given these false assumptions, it would have seemed superfluous to supply information about the route of administration. It would be like handing someone a full plate of soup and saying 'use a spoon'.

Is This Organizational Explanation Sufficient?

The system model of human fallibility is strongly endorsed by a number of high-level reports. But is that really the end of the story? The answer depends on what remedial actions are likely to be set in train as the result of adopting a wholly 'organizational accident' interpretation.

Clearly, blaming and punishing the junior doctors involved would do little or nothing to prevent the recurrence of such a tragedy; indeed, it is likely to be counter-productive. The message for healthcare institutions from this organizational analysis is clear: they should review their defences regularly in order to remedy whatever gaps may exist or could be anticipated. This would certainly be a very positive outcome. But no matter how assiduously this process is carried out, not all of the latent 'pathogens' would be eliminated. Organizational safeguards can never be entirely effective. The last line of defence would continue to be the junior doctors and nurses in direct contact with the patient. And they are unlikely to be markedly different in terms of either quality or experience from those involved in the case outlined above. We need to provide them with mental skills that help them to recognize situations of great error potential. In short, we have to make junior doctors and nurses more 'error-wise'.

Conclusions

It is evident from the case study presented here that 'organizational accidents' do occur in healthcare institutions. The identification of 'organizational accidents' enjoins us to ask how and why the safeguards failed. It also requires not only the remediation of the defective barriers, but also regular audits of all the system's defences. The same event never happens twice in exactly the same way. It is therefore necessary to consider many possible scenarios leading to patient harm. This would truly be proactive safety management, since the latent ingredients of future adverse events are already present within the organization.

Case Study 7: A GP's Prescribing Errors[9]

In analysing errors, it is customary to describe their nature and context, and then to consider possible contributing factors. The aim, at least in my business, is to frame these analyses in a way that leads to enhanced prevention – particularly when the errors have damaging consequences, as in this case. Although errors are frequently considered as the result of wayward mental processes in the heads of those at the sharp end, it is more usually the case that the contributing factors arise within the workplace, the organization and/or the system at large. It is my opinion that all of these latter factors played a significant part in causing the errors considered in this instance.

---

9   Reason, J. (2013). Proof of Evidence in Defence of the GP. Unpublished.

On the Nature of Error

One of the most important findings about human fallibility is that the same (or similar) situations create the same or similar errors in different people. In short, we are talking mainly about error-provoking situations rather than error-prone individuals. These recurrences are called error traps. The existence of an error trap does not mean that everyone falls prey to them – only that many people succumb to their powerful spell. I believe that Dr X was the victim of such a trap.

Dr X's Errors

There were two principal errors, both repeated on two different patients: an extremely unfortunate but understandable repetition given that the prescriptions for the two patients were created sequentially and in close temporal proximity. And in both cases, Dr X was prescribing remotely for patients who were not physically present: they were both elderly and in a nursing home. Both were complaining of intense ulcerous pain that was not alleviated by paracetamol.

Faced with need to alleviate severe pain swiftly, Dr X considered a non-steroid anti-inflammatory, but rejected it in favour of Oramorph at a low dose, or some (other oral) morphine at a low dose. He judged that there was too great a risk of gastric bleeding if he used the anti-inflammatory, particularly in an elderly patient. It was a considered decision likely to bring about the desired outcome. Thus, I don't regard it to have been a mistake.

He accessed Mr A's record on the computer screen. He then clicked on the icon to bring up the prescribing screen. His intention was to prescribe 2.5 mls of Oramorph, to be used four times a day.

Having entered Oramorph, the screen then brought up a drop-down menu (pick list) of various opiates. He intended to click on Oramorph, but actually clicked on the first item in the list. This turned out to be Morphine Sulphate, a concentrated oral solution, at a concentration of 20 mgs per ml. He was unaware of this, his first error.

Another page appeared automatically on the screen, enabling him to type the dosage, which he did as 2.5 mls of the solution at a maximum of four doses per day. He was not alerted to the fact that he had selected a morphine solution with a 20:1 ratio when he had intended 2:1.

He printed off the prescription and signed it, not detecting the error. Signing off on the prescription was his second error.

He then brought up the second patient's records (Mrs B) on the screen and went through the same process in producing the prescription, repeating precisely the same slips.

An explanation of these two errors implicates two issues: first, the problems associated with drop-down menus; and, second, the limited 'depth of processing' problem frequently associated with automation through computer use. I will deal with each of these separately

Drop-Down (Pick List) User Problems

Shortly after being introduced to the case, I visited Chicago to give a series of lectures and have discussions with the senior management of the Joint Commission (TJC), whose mission is to evaluate healthcare organizations for the purposes of accreditation. During one such discussion, I outlined the case above and was surprised by the reaction of the JTC's Chief Medical Officer. In a later email, she described her response as 'visceral' because drop-down menu problems were so common in the US and medication issues were their single most frequent adverse event.

My first step on returning home was to Google the following search terms. The numbers indicate the 'hits' associated with each set of terms:

- Drop-down menus medication errors   524,000 hits
- Drop-down menus GP prescribing errors   90,000 hits
- Drop-down menus errors with opiates   51,000 hits

I am well aware that not all of these 'hits' are directly relevant to this case, but if even only 20 per cent are, then the problem is huge and widespread. At the very least, these hits strongly suggest that pick lists are highly error-provoking devices – in short, they are powerful 'error traps'.

One article yielded by these searches was entitled 'The Problem with Drop-Down Menu Navigation'.[10] Some of the major user problems are listed below:

- It is easy to hit the wrong link with a single mouse click (two clicks better than one).
- The menu is hidden from the user.
- They are hard to use.
- They are not scalable.
- Some of the top-performing websites do not use them.
- Intermediate index pages are absent and navigation OS difficult.
- It is easy to get lost.

That these problems have found their way into computer prescribing is evident from the following studies drawn from a variety of web-based sources.

A study investigating problems associated with computerized provider order entry (US acronym CPOE) identified nine types of unintended consequences. Of these, the one below is the most relevant to the present case:

> *New Kinds of Errors*: CPOE tends to generate new kinds of errors such as juxtaposition errors in which clinicians click on the adjacent patient name or medication from a list and inadvertently enter the wrong order.[11]

A paediatric study[12] of CPOE medication errors found that mis-selections from drop-down menus accounted for around 10 per cent of all errors observed.

---

10   Kelly, B. (2007). Available at: http:www.zeald.com/Blog/x_post/the-problem-with-drop-down-menu-navigation.html.
11   Ash, J.S. et al. (2007). 'The Extent and Importance of Unintended Consequences Related to Computerized Provider Order Entry'. *Journal of American Medical Informatics Association*, 14(4), pp. 415–23, at p. 416.
12   Walsh, K.E. (2006). 'Medication Errors Related to Computerized Order Entry for Children'. *Pedeatrics*, 118(5), pp. 1872–9.

Other investigators[13] reviewing a wide range of publications identified six major design defects impacting on the reliable use of CPOE systems. Pick lists and drop-down menus were ranked as third.

I could offer many more findings, but I think the main point has been made clear: drop-down menus on computer prescribing systems create large numbers of wrong medication, wrong patient, and other unintended consequences. As such, they constitute well-documented error traps, involving juxtaposition errors in particular.

I do not know how close Oramorph was to the wrongly selected Morphine Sulphate on the pick list, but even if they were not close neighbours, there were likely to be other factors contributing to Dr X's errors. Mr A and Mrs B were not scheduled patients on his surgery list – their prescriptions were prompted by a phone call from a nursing home nurse whom he only knew as Jane. Preoccupation and pressure of time can induce absent-minded slips. Thus, when the word 'Morphine' appeared as the first item on the pick list, it may have elicited a 'coiled spring' reaction. The only response needed was a single click on the mouse, something requiring a small movement of a poised forefinger. At no subsequent point in Dr X's prescribing session was he alerted to his error.

Given that you don't have to be a genius to anticipate the primacy of the top list item, it seems quite remarkable to me that the designers of the system had allowed this potentially (and actually) lethal drug and dose to head the drop-down menu. Clearly the programmers were not medically trained or were themselves distracted or preoccupied. I believe that Dr X was the victim of a latent failure within the computer system that was going to catch somebody out sooner or later.

Another thing that convinces me that this was an absent-minded slip – something to which all of us are prey – was that it fell into a recognisable type. Although no slip is exactly the same as another, the various slip categories share many formal characteristics.[14] This was a recognition failure leading to the selection of a wrong object or item. An everyday example of this has occurred to me

13   Khajouei, R. (2008). 'CPOE System Design Defects and their Qualitative Effect on Usability'. *Studies in Health Technology and Informatics*, 136, pp. 309–14.

14   See Reason, *Human Error*.

more than once. I have occasionally bought shaving cream in a squeezable tube. On several occasions, I have put shaving cream on my toothbrush. Many factors conspire to produce this error:

  a. performing a highly routine activity with my mind on other things;
  b. proximity – the toothpaste and the shaving cream are kept together on a shelf close to my right hand;
  c. similarity – both objects are very alike in shape and size; and
  d. expectancy – I expected that what I had grasped was toothpaste rather than shaving cream.

This brings me to another important point: outcome bias. There is a strong inclination, particularly among children and lawyers, to believe that errors having dire consequences arise from particularly reprehensible mental processes. In other words, they presume symmetry between consequence and cause. My shaving cream slip left me blowing bubbles and feeling foolish, but it was essentially inconsequential. Dr X's errors were not. However, it's the circumstances rather than the prior mental processes that largely determine the consequences. I could, for example, switch on the toaster rather than the kettle. The result would be mild annoyance. But if I had made a comparable error on the flight deck of an aircraft or in the control room of a nuclear power plant, the result could have been catastrophic. Such a slip contributed to the Chernobyl disaster. Yet, as always with human failures in complex systems, errors hardly ever arise from single causes; they are more often the end product of a concatenating cascade of latent failures in the workplace and the system.

I have already touched upon some of the design defects present within Dr X's computer system. But how did it get selected and bought? These are systemic rather than end-user issues. The decision was probably made by the practice managers. How many other GPs have experienced comparable (though probably less harmful) errors in using their practice computers? One of the unhappy features of healthcare is that little slips can cause big disasters. Patients are vulnerable. Healthcare, by its nature, is highly error-provoking. Yet doctors, by and large, know little or nothing about the nature and varieties of human error.

And, unlike aviators, they equate errors with incompetence, stigmatizing, marginalizing and sanctioning the error-makers. As a result, they miss the golden opportunity of treating errors as valuable learning experiences – unlike others who operate hazardous technologies. By isolating the fallible person from his or her context, they also lose the opportunity of scrutinizing the system at large for contributing latent failures.[15]

Failing to Detect the Error on Signature

Why did Dr X not detect the prescription error on the two occasions when he signed the prescriptions? I can only offer two possibilities. First, he didn't see the error because he had no reason to expect it, having had no warning alerts from the computer system. Expectations play a very large part in perception: you often falsely perceive what you strongly expect and, by the same token, fail to detect things that you don't expect to be there. Dr X expected the computer to have got it right.

The second point also relates to the dark side of computer usage. Every psychology undergraduate learns about the 'levels of processing' effect.[16] Contrary to the then existing orthodoxy that claimed separate stores for short-and long-term memory, Craik and Lockhart argued that the likelihood of correct recall was a function of the depth of mental processing. Shallow processing based upon phonemic and orthographic components leads to a fragile or non-existent memory trace; deep processing involving semantic memory and self-referencing results in a more durable memory. It has been found in many contexts that computer-based automation denies those involved of this deeper processing. It is highly unlikely, for example, that the errors would have been made had they been written in the old-fashioned way and, if they had, they would have stood a much better chance of detection. In summary, shallow, computer-related processing leads to both poor attention and very limited recall.

---

15   Reason, J. (1997). *Managing the Risks of Organizational Accidents*. Aldershot: Ashgate.

16   Craik, F.I.M. and Lockhart, R.S. (1972). 'Levels of Processing: A Framework for Memory Research'. *Journal of Verbal Learning and Verbal Behavior*, 11, pp. 671–84.

From the mid-1970s onwards, when cheap and powerful computational devices became widely available, there was a rush to automate, partly in the belief that it would eliminate the problem of error by taking the human out of the loop. But whilst the modern generation of computer systems have greatly reduced certain kinds of errors (usually fairly trivial ones), they have introduced new and exceedingly dangerous error forms. Error has not been removed, it has been relocated. The British engineering psychologist Lisanne Bainbridge summed up these problems in the elegant phrase 'the ironies of automation'.[17]

Conclusion

Contrary to my first impressions, I do not find Dr X grossly negligent. He fell into a well-documented and dangerous error trap. If there is any 'grossness' to be found in this unhappy affair, it relates to the computer system sitting on Dr X's desk. This failed him in at least two ways:

- It encouraged the commission of a lethal error by putting Morphine Sulphate at the top of the drop-down menu.
- It subsequently failed to alert him to this error.

There were three victims in this case: Mr A, Mrs B and Dr X.

Case Study 8: Texas City Refinery Explosion 2005

Background

In March 2005, a hydrocarbon vapour cloud exploded at BP's Texas City refinery, killing 15 workers and injuring over 170 others. The site was designed to convert low octane hydrocarbons, through various processes, into higher octane hydrocarbons that could be blended into unleaded petrol. This was the second-largest oil refinery in Texas and the third-largest in the US. BP acquired the Texas City refinery as part of its merger with Amoco in 1999.

---

17 Bainbridge, L. (1987). 'Ironies of Automation'. In J. Rasmussen et al. (eds), *New Technology and Human Error*. Chichester: Wiley.

The refinery was built in 1934, but had been badly maintained for several years. A consulting firm had examined conditions at the plant. It released its report in January 2005, which found many safety issues that included 'broken alarms, thinned pipe, chunks of falling concrete, bolts falling 60ft and staff being overcome with fumes'. The refinery had had five managers since BP had inherited it in its 1999 merger.

I visited a BP site in Sale Victoria (Australia) in 2001 and was impressed by its various safety management systems. They looked very sophisticated on paper in their fat ring binders. They included a Management of Change (MOC) process, a BP Pre-Startup Safety Review (PSSR) and (see more later) an impressive overriding safety process called the Operating Management System (OMS). I came away thinking 'these are the good guys'.

After remedial work had been completed on the 170 ft raffinate tower (used for separating lighter hydrocarbon components), the PSSR was conducted. Its purpose was to establish that safety checks had been carried out and that all non-essential personnel were clear during the start-up. Once completed, the PSSR would be signed off by senior managers, but these essential safety procedures were not completed. These included an inoperative pressure control valve, a defective high-level alarm and a defective sight tower-level transmitter that had not been calibrated. It was also discovered that one of the trailers, used to accommodate contractors, was too close to the process and thus liable to severe damage in the event of an explosion.

The explosion occurred on 23 March 2005. According to BP's own accident report, the direct cause was 'heavier than air hydrocarbon vapours combusting after coming into contact with the ignition source, probably a running vehicle engine'. The hydrocarbons came from the liquid overflow from the blow down stack following the activation of the splitter tower's over-pressure protection system. This, in turn, was due to the overfilling and overheating of the tower contents.

Both the BP and the Chemical Safety and Hazard Investigation Board reports identified numerous technical and organizational failures at the refinery and within corporate BP. These organizational failings included:

- Corporate cost-cutting.
- A failure to invest in the plant's infrastructure.
- A lack of corporate oversight on both safety culture and accident investigation programmes.
- A focus on occupational safety (lost time injuries) and not process safety.
- A defective management of change process.
- The inadequate training of operators.
- A lack of competent supervision for start-up operations.
- Poor communications.
- The use of outdated and ineffective work procedures.

Technical failures included the following:

- A too-small blowdown drum.
- A lack of preventative maintenance on safety-critical systems such as imperative alarms and level sensors.
- The continued use of outdated blowdown drum and stack technology when replacements had been feasible alternatives for several years.

BP was charged with criminal violations of federal environmental laws and was named in several lawsuits from victims' families. The Occupational Safety and Health Administration (OSHA) gave BP a record fine for hundreds of safety violations. In 2009, the OSHA imposed an even larger fine after claiming that BP had failed to implement safety improvements after the disaster. In 2011, BP announced it was selling the refinery in order to pay for ongoing compensation claims and remedial activities following the Deepwater Horizon disaster in 2010. This will be discussed in the next case study.

Case Study 9: Deepwater Horizon in the Gulf of Mexico[18]

Deepwater Horizon was a semi-submersible, mobile, floating, dynamically positioned oil rig, operated mainly by BP and, at the time of the explosion, was situated in the Macondo Prospect roughly 41 miles off the Louisiana coast. The disaster was in two

18   *BP Fatal Accident Investigation Report.* Texas City, Texas, 2005.

parts: the explosion on 20 April 2010, killing 11 people, and an oil spill discovered on 22 April – the spillage continued for 87 days without pause and covered an area of the Gulf equivalent to the size of Oklahoma. This was the largest accidental oil spill in the story of petroleum exploration. After several failed efforts, the well was finally sealed on 19 September.

Among others, two companies were involved in the event: Transocean, which owned/operated the drilling rig, and Halliburton, which carried out (among other things) essential cementing work. BP (with a 65 per cent share) was in overall charge. Many problems arose because of conflicts between these three companies.

Deepwater oil exploration is highly complex – made more so by earlier attempts to abandon the well. It is best therefore that I confine myself to the salient points. These have been admirably summarized by Professor Patrick Hudson, who served as an expert witness on behalf of Halliburton. The main points of the accident sequence are listed below:

- At 9.45 am CDT on 20 April 2010, high-pressure methane gas from the well expanded into the drilling riser (it was underbalanced with seawater) and rose into the drilling rig, where it ignited and exploded, engulfing the platform.
- The Blow Out Preventer was activated late and failed to shut in the well.
- The riser was not disconnected and escaping gas ignited. As a result, 11 people died.
- The drilling rig sank after two days.

On 8 March, just over a month before the explosion, there was what Patrick Hudson termed a 'dress rehearsal'. There was an influx of hydrocarbon, what the drillers term a 'kick'. It was detected late – the mud-logger was not believed. Nonetheless, the well was shut in successfully. The costs including rig time exceeded $10 million. No report was made to the London office.

Halliburton was subsequently engaged in cementing the well. Its purpose was to centralize the production casing to avoid channelling. Centralizers were necessary to ensure that the annulus was constant, otherwise the cement leave's mud behind.

This allowed the possibility of hydrocarbon escaping through the mud. Halliburton's software recommended 21 centralizers. BP unilaterally decided to use only six.

The well needed to be cleaned by circulating mud. BP rejected Halliburton's recommendation to run a full bottom-up clean. It was also necessary to carry out both positive and negative pressure tests to ensure that the cement barrier was effective. This procedure also encountered a variety of BP cost-saving measures.

In 1990, John Browne was appointed as CEO of BP Exploration and Production upstream. Five years later, he was appointed CEO of BP. He established a rigorous discipline cost-cutting. This was judged to be a major cause of the problems at Texas City which had 2.25 per cent across the board cost reduction targets prior to the disaster. The financiers loved him. He created a community that loss-averse, unlike its risk-averse competitors (Shell and Exxon-Mobil). His successor, Tony Hayward, continued the cost-cutting culture with the maxim 'Every dollar counts'.

Patrick Hudson's conclusions with regard to the explosion and oil spill:

- The accident was preventable.
- If BP's OMS had been applied rigorously, the temporary well abandonment could have been completed safely.
- The dominant failures were associated with no risk analyses or assessments despite major changes and problems with the well.
- These primary causes could be related back to BP's organizational safety culture. In particular, the problem was – as noted at Texas City – that safety was seen as personal rather than personal plus process safety.

In November 2012, BP pleaded guilty to 11 counts of manslaughter, two misdemeanors and a felony count of lying to Congress. The Environmental Protection Agency announced that BP would be temporarily banned from new contracts with the US government. BP and the Department of Justice agreed to a record-setting $4.525 billion in fines and other payments. As of February 2013, criminal and civil settlements had cost BP $42.2 billion. Further legal proceedings are not expected to conclude until 2014 are currently ongoing to determine payouts and fines under the Clean Water Act and the Natural Resources Damage Assessment.[19]

Case Study 10: Loss of Air France 447 Off the Brazilian Coast

Background

On 1 June 2009, Air France Flight 447 (AF 447) took off from Rio de Janeiro with 228 passengers and crew on board. It was scheduled to arrive at Charles de Gaulle Airport, Paris, after an 11-hour flight, but it never arrived. Three hours later, it was at the bottom of the ocean. There were no survivors. Of all the case studies considered here, I find this one the most horrific, mainly because, unlike most of the others, the victims could be anyone. At three hours into a long-distance flight, it is likely that dinner had been served and that most of the passengers would have been composing themselves for sleep, reading or watching videos.

The flight crew comprised the captain and two co-pilots. The aircraft was an Airbus 330-200. In line with European Commission Regulations, the captain told the two co-pilots that he was going to the bunk bed, just aft of the flight deck for a scheduled rest. Unfortunately, the captain did not give explicit instructions as to who should be the pilot flying (PF). The co-pilot he left with primary controls was the junior and least experienced of the two, though he did have a valid commercial pilot's licence. This omission had serious consequences because the captain should have left clear instructions with regard to task sharing in his absence from the flight deck.

The Event

In the early hours of 1 June 2009, the Airbus 330-200 disappeared in mid-ocean beyond radar coverage and in darkness. It took Air France six hours to concede its loss and for several days there was absolutely no trace, and even when the wreckage was eventually discovered, the tragedy remained perplexing. How could a highly automated state-of-the-art airliner fall out of the sky?

---

19   See https://en.wikipedia.org/wiki/Deepwater_Horizon_oil_spill.

It was five days before debris and the first bodies were recovered. Almost two years later, robot submarines located the aircraft's flight recorders.

The immediate cause of the crash appeared to be a failure of the plane's pitot tubes – forward-facing ducts that use airflow to measure airspeed. The most likely explanation was that all three of the pitot probes froze up suddenly in a tropical storm. This freezing was not understood at the time.

The flight crew knew that there were some weather disturbances in their flight path. But because of the limited ability of their onboard weather radar, they didn't realize that they were flying into a large Atlantic storm. The danger with such storms was that they contained super pure water at very low temperatures. On contact with any foreign surface, it immediately turns to ice. This it did on contact with the pitot probes, overwhelming their heating elements and causing the airspeed measurements to rapidly decrease. This, in turn, caused the autopilot to fail, requiring the pilots to take control of the aircraft manually. The disengagement of the automatic systems created an overload on the Aircraft Communications Addressing and Reporting System and the Electronic Centralized Aircraft Monitoring messages. The pilots were not equipped to cope with this cascade of failure warnings and control the aircraft. What could have saved them? The by-the-book procedure is to set the throttles to 85 per cent thrust and raise the elevators to a five-degree pitch.

Unfortunately, the PF repeatedly made nose-up inputs that exacerbated the failures, putting the aircraft into a stalled condition when there had already been two stall warnings. He did this by pulling back on the stick – unlike a more conventional control column or yoke, the Airbus stick was a short vertical control (located to the right of his seat) that responded to pressure, but did not do so visibly. Without this visual cue, the pilot not flying (PNF) took half a minute or so to figure out what the PF was doing. The PNF said 'we're going up, so descend'. Seconds later, he ordered 'descend' again. At this point, the aircraft was in a steep nose-up attitude and falling towards the ocean at 11,000 ft per second. During that time, there were over 70 stall warnings. The PF continued to pull back on the stick and continued to climb at the maximum rate.

Seconds after emergency began, the captain returned to the flight deck. He sat behind the two co-pilots and scanned the instrument panel, and was as perplexed as his colleagues. There is one final interchange:

> 02:13:40   (PNF) 'Climb … climb … climb … climb'
> 02:13:40   (PF) 'But I've had the stick back the whole time!'
> 02:13:42   (Captain) 'No. No. No … Don't climb … No. No'.
> 02:13:43   (PNF) 'Descend … Give me the controls … Give me the controls'.
> 02:14:23   (PNF) 'Damn it, we're going to crash … This can't be happening'.
> 02:14:25   (PF) 'But what's going on?'

Conclusion

When we study a group of individuals, we recognise the commonalities and note the differences, and in the case of people it is usually the latter that are regarded as the most interesting and important. But for the case studies outlined above, it is the other way round. We readily appreciate the differences in domain, people and circumstances; but what interests us most are the common features, particularly of the systems involved and of the actions of the participating players.

One obvious feature of these case studies is that they all had bad outcomes. Of more interest, though, is the way that these grisly things happened. It was thoughts like these that set the Swiss Cheese Model in motion. In my view, all the events had two important similarities: systemic weaknesses and active failures on the part of the people involved. Neither one is a unique property of any one domain. The question this chapter addressed was how do these two factors combine to bring about the unhappy endings? In most cases, the answers lie in what was happening on the day and in the prior period. These are almost too varied to be sensibly categorised, but that is what this book is about. But, more than that, it is part of an effort to discover and predict the ways that these malevolencies combine to achieve their mischief.

Chapter 7
Foresight Training

Error Wisdom

This chapter focuses on the healthcare domain. Although organizational accidents have their origins in system weaknesses, my argument here is that purely systemic counter-measures are not enough to prevent tragedies. This is especially the case in healthcare, when the last line of defence is a doctor or nurse working in close proximity to the patient. How can we make these 'sharp-enders' more error-wise and more alert to the risks and hazards present within the system? In short, how can we make them more 'mindful' of the possible dangers?

I was led to this line of thought by the tragic case of Wayne Jowett, an 18-year-old boy who was largely recovered from acute lymphoblastic leukaemia (the case history is discussed in Chapter 6 – Case Study 6). He mistakenly received an intrathecal (spinal) injection of the cytotoxic drug vincristine and died three weeks later – vincristine destroys neural tissue. It happened at the prestigious Queen's Medical Centre in Nottingham in January 2001. It was the subject of an in-depth study by Professor Brian Toft,[1] an expert on accident analysis. It is rare that such medical incidents receive close attention and the report made available in the public domain. It would be useful if you could look back at Case Study 6 to familiarize yourselves with its details.

Vincristine accidents of this nature have happened many times before. Wrong-route administration of drugs is a relatively common medical incident. The Nottingham hospital was well aware of these dangers; indeed, it had protocols in place to protect

---

1   Toft, B. (2001). *External Inquiry into Adverse Medical Incident Occurring at Queen's Medical Centre*. London: Department of Health.

against this error that went beyond the national requirements. One of these was that intravenous and intrathecal drugs should be delivered to the ward in separate packages and administered on different days (see Chapter 6).

I think this may have been due to what I have rather pompously called 'the lethal convergence of benevolence'. Healthcarers, by and large, care very much about the welfare of their patients. Wayne Jowett was a poor attender at the day ward. He had missed an appointment that morning and had turned up unannounced with his grandmother at 4.00 pm. It is my hunch that the pharmacists and the nurses colluded to bypass the protocols and to make sure that his medication would be ready for him when he turned up. Both were held in the ward refrigerator in a clear plastic packet. Contrary to the protocol, both drugs would be given at the same time – on the reasonable assumption, I believe, that it would be difficult to get Wayne to attend on two separate days. If the intravenous and intrathecal drugs had been in separate bags, then it would not have been necessary for the specialist registrar to supervise the process in the case of the former.

Mental Skills to Enhance Risk Awareness

What would it take to make the alarm bells ring before the fatal actions occur? What would it take to step back and ask for help?

The situation in the Queen's Medical Centre ward at 4.00 pm already possessed some ominous indications. The SpR had only been on the job for less than three days. He was instructed to do no clinical work, but was required to shadow a locum consultant – who was a part-timer and was not available. The ward sister had gone home. And the junior doctor whom he was to supervise in giving spinal injections knew little or nothing about the dangers of vincristine or about the patient. The SpR assumed otherwise. He was only trying to be helpful. Let us look briefly at some other domains.

A defining feature of high-reliability organizations (HROs)[2] is a preoccupation with the possibility of failure, both human and technical. They expect errors and breakdowns to occur and are

2   Weick, K.E. (1995). *Sensemaking in Organizations*. Thousand Oaks: Sage.

prepared to deal with them. In short, they are characterized by chronic unease and an intelligent wariness.

Demonstrations of how the mental skills underlying risk awareness can be implemented exist in a number of hazardous industries.[3] One example is Esso's 'step back five by five' programme. Before starting, the worker is enjoined to take five steps back metaphorically and take five minutes to think about what might go wrong and what needs to be done to prevent it. But it is not enough to exhort front-line workers to be more vigilant – these mental skills need organizational support. The Western Mining Company of Australia has appreciated this need. It has instituted a programme called 'take time, take charge' that aims to get people to stop and think about the possible risks and then take some appropriate action when this is necessary. What makes this work, however, is the organizational back-up. Each day, supervisors ask workers for 'take time, take charge' instances where corrective action was required. These instances are reported to weekly meetings with managers, and the organizational actions are fed back to the workers. A crucial feature of this programme is that there is a full-time senior manager whose sole responsibility is to supervise and facilitate the process on a company-wide basis. In short, individual mindfulness of danger needs to be sustained and supported by a collective mindfulness of the operational risks.

Heightened risk awareness is something that many healthcare professionals acquire as a result of long experience in the field. There is considerable evidence to show that mental preparedness – over and above the necessary technical skills – plays a major part in the attainment of excellence in both athletics and surgery. Could we shortcut the experiential learning process by providing front-liners and often junior healthcarers with training in identifying high-risk situations?

A case for providing such training could be made from many areas of healthcare, but few are more compelling than that derived from the analysis of transfusion errors. Haemovigilance schemes such as the UK's Serious Hazards of Transfusion

---

3 See Hopkins, A. (2007). *Lessons from Gretley: Mindful Leadership and the Law.* Sydney: CCH.

(SHOT) provide an insight into the errors and system failures leading to 'wrong blood to patient' incidents.[4] SHOT now has data from seven years of adverse and near-miss reporting. These data show that there are several weak links in the transfusion chain – request and prescription errors, blood sampling errors in the hospital, laboratory errors, and errors in blood collection and administration. From the present perspective, however, the most important finding is that 87 per cent could have been detected at the bedside. The key point is that no matter where in the transfusion chain the problem occurs, detection and recovery by the nurse at the bedside could thwart an adverse event. And the same is true of other organizational accidents in healthcare. The point of delivery is the last line of defence.

The Three-Bucket Model

When we were children, adults would frequently tell us to stop and think – good advice, except that they rarely told us what to think about. I have sought to overcome this omission with the 'three-bucket' model described below. This forms the basis of the mental skills designed to impart error wisdom and risk awareness and is illustrated in Figure 7.1.

Figure 7.1   The three-bucket model for assessing risky situations

4   Stainsby, D. (2005). *Errors in Transfusion Medicine*. Newcastle-upon-Tyne: National Blood Service.

The buckets correspond to the three features of a situation that influence the likelihood of an error being made. One bucket reflects the well-being or otherwise of the front-line individual; the second relates to the error-provoking features of the situation; and the third concerns the nature of the task – individual tasks and task steps vary widely in their error potential.

In any given situation, the probability of unsafe acts being committed is a function of the amount of 'bad stuff' in all three buckets. Full buckets do not guarantee the occurrence of an error, nor do nearly empty ones ensure safety (they are never completely empty) – we are dealing with probabilities rather than certainties. The contents of each bucket are scaled from one to three, summing to nine in total.

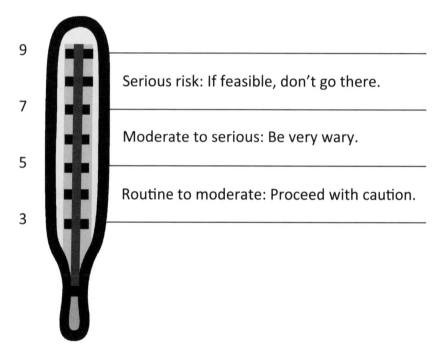

Figure 7.2   How the buckets might be 'read' by junior staff working alone

People are very good at making rapid intuitive ratings of situational features. Together with some relatively inexpensive instruction on error-provoking conditions, front-line carers could acquire the mental skills necessary to make a rough-and-ready assessment of the error likelihood in any given situation. Subjective ratings totalling between six and nine should set alarm bells ringing. Where possible, this should lead them to step back and seek help.

The three-bucket model seeks to emphasize the following aspects of mental preparedness:

- Accept that errors can and will happen.
- Assess the local 'bad stuff' before embarking on a task.
- Be prepared to seek more qualified assistance.
- Don't let professional courtesy get in the way of establishing your immediate colleagues' knowledge and experience with regard to the patient, particularly when they are strangers.
- Understand that the path to adverse events is paved with false assumptions.

As mentioned earlier, individual mindfulness of the local risks is not enough; it needs to be informed, supported and sustained by collective mindfulness. Figure 7.3 shows one way in which these two forms of risk mindfulness can be combined to enhance system resilience.

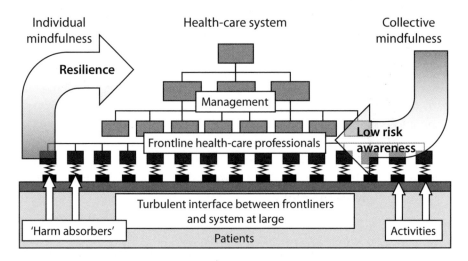

Figure 7.3   Balancing the person and the system

This figure focuses upon the turbulent interface between the healthcare organization (as represented by the organogram) and patients. The organogram shows successive levels of the system from top management to healthcare professionals on the front line. The zigzag lines beneath each professional – analogous to shock absorbers – represent the mental skills and techniques for identifying and avoiding error-provoking situations. These aspects of 'error wisdom' are called harm absorbers. Their combined influence feeds back up the organization as individual mindfulness. A comparable downward arrow – termed collective mindfulness – emphasizes the necessity for top-down support. This takes many forms: training front-line staff in error awareness, error detection and error recovery, and then monitoring and refreshing these mental skills. Most importantly, it entails a cultural change that empowers staff to step back from risky situations and seek help – where this is feasible.

Foresight Training

So far, we have considered the nature of error wisdom, a mental state that involves an enhanced awareness of the possible dangers and opportunities for going wrong embodied in a particular situation. We have been discussing healthcare, but the principles apply in any hazardous domain. For the remainder of this chapter, we go beyond mere awareness and consider the ways and means by which we can hope to convert it into a lasting mental skill. Like all skills, it needs a considered course of instruction followed by continual reinforcement and practice. These processes have been termed foresight training. This label is owed to senior researchers at the National Patient Safety Agency (NPSA), now sadly defunct (another victim of the Department of Health's and its political masters' insatiable hunger for tweaking), but not before they had designed, trialled and disseminated an excellent foresight training programme, originally intended for nurses.

Before outlining the main steps of the NPSA's Foresight Training Programme (a detailed account is available on the NPSA's website),[5] I would like to mention a paper written by three

---

5   See http://www.npsa.nhs.uk.

senior members of the NPSA's research staff – Dr Jane Carthey, Sandra Meadows and Dr Maureen Baker, now chairperson of the Royal College of General Practitioners. Jane Carthey was the first author and she was also the main human factors observer on the large-scale study of the arterial switch operation (ASO). It involved 21 UK cardiothoracic surgeons in 16 centres over a period of 18 months. These case studies provide an excellent picture of what went right as well what went wrong in the ASO procedures.

The ASO is performed on babies who are born with the great vessels of the heart connected to the wrong ventricle. The aorta is connected to the right ventricle and the pulmonary artery to the left ventricle. Without surgical intervention, the babies would die. The procedure involves putting the patient on to a heart-lung bypass and freezing the heart. The machine provides mechanical support to the heart and the lung. Once the heart is frozen, the surgeon transects the native aorta and excises the coronary arteries from the native aorta and re-implants them into a neo-aorta. Given that coronary arteries are only millimetres in width and very fragile, the surgeon is working at the edge of human capability. Dr Carthey gives three examples of using foresight to avoid an incident.

Case Study 1: Foreseeing Likely Failure

Most patients have a Type A coronary arterial pattern (around 80 per cent). But in rare cases, patients are born with an intramural coronary pattern (Type B), for which it is technically very difficult to achieve a good surgical repair.

Surgeon A, to whom the patient had been referred, was informed by the cardiologist that the coronary arterial pattern was likely to be Type B. The surgeon had carried out fewer than 10 switch operations per year and had never successfully repaired a Type B anatomy. Both of his senior surgical registrars were similarly inexperienced. This centre also lacked crucial hardware that could do the work of the heart and lungs for several weeks while the organs recover. Surgeon A was aware that there was another paediatric cardiothoracic surgeon (Surgeon B) in the UK who was highly experienced in switch operations and had good results with intramural cases. The baby was transferred to his unit accompanied by Surgeon A and one of the surgical registrars who observed Surgeon B operating. The case went smoothly and there were no post-operative problems.

This was a good example of the three-bucket model in action. Surgeon A appraised himself, the context and the task in a way that optimized patient safety. Transferring a patient to another unit for surgery is not an easy decision to make in a macho culture that held the view that a good surgeon should be able to cope with anything.

Case Study 2: Using Foresight to Recover from an Incident

Due to non-surgical interruptions, just prior to the patient being put on the heart-lung machine, the surgeon failed to initiate the protocol in which the anaesthetist administers heparin, an anti-coagulant drug, prior to going on the heart-lung machine. The protocol requires that the anaesthetist confirms the administration twice before the perfusionist starts the bypass procedure. In this case, the anaesthetist had maintained a close awareness of the surgical steps and alerted the surgeon that he had omitted the heparin protocol. This was rectified and the operation proceeded normally. This shows that a key element of foresight is keeping abreast of bigger picture rather than focusing on one's own specialized activities. By maintaining situational awareness, it is possible to foresee and recover from an incident.

Case Study 3: Having Contingencies in Place

It is sometimes not enough simply to foresee a problem – it is also necessary to have contingencies in place to deal with these foreseeable difficulties. This is illustrated by the surgeon who went back on to bypass six times to correct problems. Each time, he was testing a different hypothesis. In the end, the baby was saved. This was remarkable in two respects: first, surgeons hate going back on bypass; second, this surgeon also tested a range of solutions. These are indicators of a high degree of surgical excellence. A more common response is to become fixated on one possible remedy and to persist with it.

The NPSA's Foresight Training Programme

Using the three-bucket model as its basis, the NPSA developed a Foresight Training Programme in 2008.[6] It was aimed principally at nurses and midwives working in primary care, acute care and mental healthcare settings. The programme comprised video and written scenarios that can be used to raise awareness of potential harm. The principal objectives of foresight training are listed below:

- To improve awareness of the factors that increase the likelihood of patient safety incidents and to enhance empowerment among staff to intervene in order to forestall harm.
- To improve understanding of 'risk-prone situations'.
- To help healthcare professionals make the necessary behavioural changes to adopt a more proactive approach to managing the safety of complex dynamic healthcare systems.

More on the Three Buckets

The *self bucket* relates to personal factors: level of knowledge, level of skill, level of experience, current well-being or otherwise. For example:

- a healthcare professional working at night is more likely to be affected by the consequences of shift work, such as fatigue;
- someone who has had a bad night's sleep due to a sick child;
- someone whose partner has just lost their job.

---

6   NPSA (2008). *Foresight Training: Resource Pack*. London: NPSA.

The *context bucket* reflects on the nature of the work environment: distractions and interruptions, shift handovers, lack of time, faulty equipment and lack of necessary materials. For example:

- a cluttered environment makes it difficult to see what equipment is available and whether it is being maintained r properly calibrated;
- a senior nurse arrives on duty having missed the handover and two members of his team are away sick;
- an agency nurse does not know the layout of the ward.

The *task bucket* depends on the error potential of the task. For example, steps near the end of an activity are more likely to be omitted. Certain tasks (e.g., programming infusion pumps) are complex and error-prone.

In order for foresight training to be disseminated as widely as possible, it is necessary to train facilitators to lead sessions. This preparation is contained in the Foresight Resource Pack (see the NPSA website). This provides an excellent summary of the whole programme.

Chapter 8
Alternative Views

Barry Turner, Nick Pidgeon and 'Man-Made Disasters'

It is, I believe, fitting to begin this survey of alternative theoretical views with Barry Turner, a sociologist at the University of Exeter, who – if he didn't actually coin the term 'organizational accident' – laid the groundwork for understanding organizational breakdown in his pioneering book *Man-Made Disasters* in 1978.[1] Later, Turner's work was updated in a second edition published in 1996 by his erstwhile research student, Nick Pidgeon.[2]

Turner based his analysis on inquiries into accidents and disasters occurring in Britain over an 11-year period. His most important concept was 'incubation'. In other words, organizational disasters develop over long periods of time – in short, they incubate within the system. Warning signs are ignored or misunderstood or even integrated into the pattern of organizational 'normalcy'. As a result, safeguards and defences either do not get built or are allowed to atrophy.

Very recently, Carl Macrae has applied Turner's thinking to healthcare systems. Macrae described healthcare incubation very eloquently as follows:

> In healthcare organizations some of the key sources of missed, miscommunicated or misinterpreted signals of risk are closed professional cultures, competing and conflicting demands, and the inherent ambiguity of many forms of adverse event.[3]

1   Turner, B. (1978). *Man-Made Disasters*. London: Wykeham.
2   Turner, B. and Pidgeon, N. (1997). *Man-Made Disasters*, 2nd edn. London: Butterworth-Heinemann.
3   Macrae, C. (2014). 'Early Warnings, Weak Signals and Learning from Healthcare Disasters'. *BMJ Quality and Safety*, published online, 5 March.

Disasters, as noted elsewhere, are immensely diverse in their surface details. But Turner and Pidgeon have identified a set of developmental stages that appear universal. These are summarized below:

Stage 1: Notionally normal starting points.
          Initial culturally accepted beliefs about hazards.
          Associated precautionary norms (laws, codes, etc.).
Stage 2: 'Incubation period' – the accumulation of unnoticed events that are at odds with beliefs and norms for avoiding hazards.
Stage 3: Precipitating event that attracts attention.
Stage 4: *Onset* – the immediate consequences of the collapse of cultural precautions become apparent.
Stage 5: *Rescue and salvage* – first-stage cultural adjustment creating ad hoc adjustments allowing for rescue and salvage.
Stage 6: *Full cultural adjustment and beliefs* are adapted to fit the newly gained understanding of the world.

These notions do not necessarily conflict with the idea of latent conditions; rather, their sociological emphasis upon cultural adjustments enriches them.

Carl Macrae[4]

As mentioned earlier, Carl Macrae has recently published a paper that adapts the Turner and Pidgeon scheme to understanding healthcare disasters such as that occurring at Mid-Staffordshire Foundation Trust. It concludes by suggesting three practical ways that healthcare organizations and their regulators can improve safety and address emerging risks:

- engage in practices that actively produce and amplify warning signs;
- work to define and continually update a set of specific fears of failure;

4   Macrae, C. (2014). *Close Calls: Managing Risk and Resilience in Airline Flight Safety*. Basingstoke: Palgrave Macmillan

- routinely uncover and publicly circulate knowledge of the systemic risks to patient safety and the measures required to manage them.

Macrae makes the telling point that, unlike aviation and many other domains, healthcare has no (or very few) public inquiries. As he puts it: 'When it comes to learning from systems-wide failures, the healthcare system is largely flying blind'.[5] He concludes his recent paper with the following statement:

> But, above all, one of the most urgent and timely issues in patient safety remains the challenge defined by Barry Turner almost four decades ago: determining 'which aspects of the current set of problems facing an organization are prudent to ignore and which should be attended to, and how an acceptable level of safety can be established as a criterion in carrying out this exercise'.[6]

It is clear that Barry Turner's pioneering work is as relevant now as it was nearly half a century ago.

David Woods and His Associates

David Woods and his associates (especially the brilliant anaesthesiologist Richard Cook) have been the giants in the field of safety research for more than 30 years. David began his working life as a researcher at Westinghouse and his early years were spent engaging with the nuclear power generation world in the US. He was then and still is a great inspiration for me. Subsequently, he set up a human factors laboratory at Ohio State University and worked in a variety of hazardous domains.

Very conveniently, the salient points of their work have been summarized in two volumes. The first appeared as a chapter in an edited collection in 2006[7] and the second as a full book in 2010.[8]

---

5 Macrae, 'Early Warnings'.
6 Ibid.
7 Woods, D.D. et al. (2006). 'Behind Human Error: Taming Complexity to Improve Patient Safety'. In P. Carayon (ed), *Handbook of Human Factors in Healthcare*. New York: Erlbaum.
8 Woods, D.D. et al. (2010) *Behind Human Error*. Farnham: Ashgate.

I liked the first version better, maybe because it was somewhat less critical about latent failure theory – even occasionally being quite complimentary.

The second version distinguishes between first and second stories about error (I like that distinction). In the first story, error is seen as a cause of failure; in the second story, it is seen as the effect of systemic vulnerabilities deeper inside the organization. First story: saying what people should have done is a satisfying way of describing failure; second story: but it does not explain why it made sense for them to do what they did. First story: telling people to be more careful will make the problem go away; second story; only by constantly seeking their vulnerabilities can organizations enhance safety.

So far, I have no arguments. The second stories echo the message of this book. I like the elegance of the labels.

However, the following harsh question remains: to where does it lead? Safety science is not only required to produce analyses, but also to indicate remedies. What remedial actions do these views suggest?

In a phrase, 'pursue second stories'. Why? You have to because first stories are biased by outcome knowledge. The hindsight bias distorts our understanding of what happened; this narrowing of focus leads to premature closure on the actions of those at the sharp end. This obscures how people and organizations work to recognize and overcome hazards; it conceals the pressures and dilemmas that drive human performance, and generates premature and incomplete conclusions that hinder the ability of organizations to learn and improve. But, in the words of the song, 'it ain't necessarily so'. I've spent a lifetime arguing otherwise, as I hope this book and its predecessors show.

So, what is the alternative? The authors insist that it is necessary to go behind the label 'human error'. You have to look under your feet in order to appreciate how people are continually struggling to anticipate pathways to failure.

In doing this, we may begin to see (and here I paraphrase from Woods et al.):[9]

---

9   Ibid.

- how workers and organizations are continually revising their approach to work in an effort to remain sensitive to the possibility of failure;
- how we and the workers are only partially aware of the current potential for failure;
- how change is creating new paths for failure and new demands on workers, and how revising their understanding of these paths is an important aspect of work on safety;
- how the strategies for coping with potential paths can be either strong and resilient or weak and brittle;
- how the culture of safety depends on remaining dynamically engaged in new assessments and avoiding stale, narrow or static representations of risk and hazard;
- how overconfident nearly everyone is that they have already anticipated the types and mechanisms of failure, and how overconfident nearly everyone is that the strategies they have devised are effective and will remain so;
- how missing side-effects of change is the most common form of failure for organizations and individuals; and
- how continual effort after success in a world of changing pressures and hazards is fundamental to creating safety.

There is much to applaud here. But, to me, it seems more like a wish list than a practical schedule for the guidance of risk professionals. And I cannot see that this concentration on the 'second story' will necessarily escape the distorting influences of the observer and hindsight biases. To make any sense of these observations, we have to track back in time; the metaphorical world under our feet does not remain static, and so the 'second story' too will suffer the distortions of looking back.

We have been aware of hindsight bias for many years, but, like any other bias, it can be laid off for. I think the Ohio State University group make too much of it. One other thing: I bitterly resist any attempt to proscribe the language of error. Safety science rests heavily on stories that capture the subtle interactions of an event, but an incident story without mention of error or individual wrong actions is a story without a beginning – or maybe an end. Either way, it leaves an unnecessary gap.

Charles Perrow: Normal Accident Theory

Like Barry Turner, Yale sociologist Charles (Chuck) Perrow was interested in the big picture of system safety. In the aftermath of the Three Mile Island accident (1979), the most scary nuclear event in the US (nothing bad actually happened, but it very nearly did so), he wrote a brilliant book called *Normal Accidents*.[10] I have commented on it at length elsewhere,[11] but here is the gist. Accidents and bad events are inevitable in nuclear power plants and other complex, tightly coupled systems – hence they are 'normal'.

After having noted that between 60 and 80 per cent of system accidents are blamed on 'operator error', Perrow went on to make the following telling comment: 'But if, as we shall see time and time again, the operator is confronted by unexpected and usually mysterious interactions among failures, saying that had he or she should have zigged instead of zagged is only possible after the fact. Before the accident no one could know what was going on and what should have been done'.[12]

One of the accompaniments of increasing automation has been that high-risk systems such as nuclear power plants and chemical process installations have become larger and more complex. This means that greater amounts of potentially dangerous materials have become concentrated in single sites under the centralized control of fewer operators.

Complexity can be described in relation to two relatively independent system properties: complexity of interaction and tightness of coupling. Systems may be more or less linear in their structure. Relatively nonlinear complex systems possess the following general characteristics:

- Components that are not linked together in a production sequence are in close proximity.

---

10   Perrow, C. (1984). *Living with High-Risk Technologies*. New York: Basic Books.
11   See Reason, J. (1990). *Human Error*. New York: Cambridge University Press.
12   Ibid, p. 9.

- Many common-mode connections (i.e., components whose failure can have multiple effects 'downstream') are present.
- There is only a limited possibility of isolating failed components.
- There are unfamiliar or unintended feedback loops.
- There are many control parameters that could potentially interact.
- Certain information about the state of the system must be obtained indirectly or inferred.
- There is only a limited understanding of some processes, particularly those involving transformations.[13]

The characteristics of a tightly coupled system are listed below:

- Processing delays are unacceptable.
- Production sequences are relatively invariant.
- There are few ways of achieving a particular goal.
- Little slack is allowable in supplies, equipment and personnel.
- Buffers and redundancies are deliberately designed into the system.[14]

Nuclear power plants are complex tightly coupled systems. Universities on the other hand are complex, loosely coupled systems

Systems have more defences against failure. A consequence of this is that systems have become more opaque to those who manage, control and maintain them. Jens Rasmussen has called this 'the fallacy of defences in depth':

> Another important implication of the 'defence in depth' philosophy is that the system does not respond actively to single faults. Consequently many errors made by staff and maintenance personnel do not directly reveal themselves by functional response from the system. Humans can operate with an extremely high level of reliability in a dynamic environment when slips and mistakes have

---

13   Adapted from ibid.
14   Adapted from ibid.

immediately visible effects. Survival when driving through Paris during rush hours depends on this fact.[15]

Jens Rasmussen: Control Theory[16]

Jens Rasmussen, a distinguished Danish control engineer, was my guide and mentor in my early error research. He was also, among many other things, the architect of the skill-rule-knowledge distinctions which I have relied very heavily on in my error classification.[17]

Jens produced a theory that is part of a cluster of views going under the general heading of control theories (see Woods et al. 2010).[18] The key verbs in these theories are migration and adaptation. For these theorists organizations are a set of dynamic adaptive processes that are focused on achieving many specific goals and hedged around by multiple constraints. The central notions are system objectives and their associated constraints. The organization's operating states are seen as being bounded by three constraints: safety, workload and economics. These are often depicted as semi-circular boundaries surrounding the operating state shown as occupying a central space that is exposed to three kinds of pressures, each operating to push the operating state away from their respective boundaries. Safety pressure works to keep the operations away from the safety boundary; efficiency pressures push away from the economic boundary, and least effort pressures work to keep the operations away from the workload boundary. As a result, the operating state is subjected to constant migrations to and away from these boundaries.

This dynamic modelling is not concerned with individual unsafe acts, errors or violations. To quote Woods et al:

---

15   Rasmussen, J. (1988). 'What Can We Learn from Human Error Reports'. In K. Duncan (ed.), *Changes in Working Life*. London: Wiley, pp. 3–4. See also Ramussen, J. (1990). 'The Role of Error in Organizing Behaviour'. *Ergonomics*, 33, pp. 10–11.

16   Rasmussen, J. (1997). 'Risk Management in a Dynamic Society a Modelling Problem'. *Safety Science*, 27, pp. 183–218.

17   Reason, J. (2013). *A Life in Error*. Farnham: Ashgate.

18   See Woods et al., *Behind Human Error*.

> The focus of control theory is not on erroneous actions or violations, but on the mechanisms that such behaviours at a higher level of functional abstraction – mechanisms that turn these behaviours into normal, acceptable and even indispensable aspects of an actual, dynamic, daily work context.[19]

Clearly I'm a greater admirer of Jens Rasmussen's work, but I jibbed at the notion of adaptation as an explanatory concept. Our differences arose from our respective training and experience. Jens was a control engineer and adaptation, which for him had a clear meaning, but for me, a psychologist, it was a bloated and overweight term. I had spent a long time researching the varieties of adaptation, so the term alone was about as useful as 'learning' and 'development' – in short, they were vacuous labels without qualification. I had spent much time (and set many exam questions) that focused on the crucial differences between sensory and perceptual adaptation. Sensory adaptation involves the continuous diminution in conscious experience when exposed to a steady-state input from a single sensory modality. Perceptual adaptation involves adjusting to a rearrangement of visual and positional senses so that with example. Sensory adaption involves a turning down of the felt gain of a continuous input: perceptual rearrangement is more complex and involves changing our mental models so that the abnormal conformity of different sensory inputs feels 'normal'.

But that's enough of my professional pickiness. Jens Rasmussen is a giant in this field. There are others who stand high: Richard Cook, David Woods, Nancy Leveson and Renee Amalberti. They generally follow the Rasmussen line, though with some variants. What they have in common is a rejection of unsafe acts as the focus of accident investigation.

If that were all, I couldn't object; like them, if it were argued that a particular blunder was the cause of the accident, I would feel very uncomfortable. Those on the front line in direct contact with the system are mostly the inheritors of bad stuff in the system rather than the instigators of it. I do not like the proscribing of terms – the field is complex enough without taking a pope-like stance on what is or is not acceptable.

---

19   Ibid., p. 75.

I have long been intrigued by the question of what is it that distinguishes systems in comparable fields that have suffered a bad event from those that have not? All such systems are riddled with latent failures – or 'conditions' as I now prefer to call them. The only distinguishing features are what happened on the day – the local actions and circumstances; in short, the immediate contributing errors and violations and what provoked them.

Erik Hollnagel: Functional Resonance Analysis Method (FRAM)

Among the alternative approaches (alternative to the arguments set out in this book and its previous edition, that is), there is none more alternative or powerful than that recently presented by Erik Hollnagel.[20] A Dane and a close friend, he is a psychologist by training, but now he is much more than that – a human factors theorist and a computer scientist – he is truly a polymath in the broad Renaissance sense. I stand in awe of his intellect and his work, but I have to confess that I do not understand it all. This failing I feel is entirely mine.

The FRAM story begins (and continues) with a tangle of 'nots' – saying what FRAM is not. There are so many of them that I'm not sure I can catch them all, but I will list the main ones below:

- FRAM, Erik insists, is a method, not a model – he even says it again in Latin: method-*sine*-model. I've spent a working lifetime tinkering with safety-related methods and models, and to me the differences are very small; they shade into one another and are two related parts of the same conceptual whole, one more theoretically orientated than the other. I think perhaps Erik recognizes this since towards the end of the book he often refers to FRAM as a model.
- FRAM, Erik argues, is proactive and not reactive. He means that FRAM is not dependent on breakdowns; it deals with successful actions, unlike the majority of other safety models. But confusingly two out of three of his worked examples of FRAM start from some kind of glitch or failure.

---

20   Hollnagel, E. (2012). *FRAM: The Functional Resonance Analysis Method.* Farnham: Ashgate.

- Erik concedes at the outset of the relevant chapter that: 'It is natural and nearly irresistible to think of events as if they develop in a step-by-step progression ... this is called linear thinking'.[21] But FRAM is not linear – neither simple nor complex. The problem arises, he argues, from confusing event succession with causality. He cites the philosopher David Hume, who stated that if event A preceded event B, it only means that they are ordered in time, not that A is the cause of B. On the face of it, that is hard to contest. It surely depends on the temporal proximity of A and B. But I am, of course, being too simple-minded.

So what, according to FRAM, causes what? The essence of FRAM is decomposition – its intellectual ancestors are Hierarchical Task Analysis (HTA) and various computer programming models. The answer to the question posed above is not straightforward: FRAM breaks down activities in a complex socio-technical system into six functions (and their associated variability), and causality arises as an emergent property of the cross-talk between them. The method is very sophisticated, although in its detailing rather than its complexity.

As an undergraduate, I learned the difference between resultant and emergent outcomes. The former could be traced to the prior causal conditions; the latter, emergent outcomes, was a shorthand way of saying that we didn't understand how they happened. But then I was an undergraduate a long time ago.

FRAM has four procedural steps:

- identifying and describing the functions;
- identifying the variability;
- determining how variability may be combined;
- how FRAM outcome can improve practice.

Each function can be described by six different aspects or features:

- *Input* – that which the function transforms or that which starts the function.

21   Ibid., p. 11.

- *Output* – that which is the result of the function.
- *Preconditions* – conditions that must exist before a function can be carried out.
- *Resources* – that which the function needs when it is carried out or consumed to produce the output.
- *Time* – temporal constraints affecting the function: start time, finishing time and duration.
- *Control* – how the function is controlled or monitored.

A FRAM function is represented graphically by a hexagon, where each corner corresponds to one of the aspects above. For the rest, I suggest that you consult Chapters 5–8 of Erik's book. The book ends with three worked examples and some afterthoughts. I have neither the space nor the inclination to go any further with FRAM, though Erik's book does a good job of explaining the details (and there are a lot of those).

Do I understand it better after two weeks of wrestling with the many details? Yes. Do I believe that FRAM is God's gift to accident analysis? For some, it will be, but for the majority, I suspect not. I should conclude by stating that FRAM has two applications: as an event analysis tool and in risk assessment. It's a powerful instrument – in the right hands. Nature has determined that I am genetically shaped to favour simplicity; however, the world of complex sociotechnical systems is anything but simple. 'Swiss cheese-type' metaphors are easily understood and disseminated – yet maybe we need to move on and FRAM could be one of the ways forward. We shall see. Homo erectus did, after all, become homo sapiens.

Chapter 9
Retrospect and Prospect

This chapter has a lot to say about patient safety. I make no apology for this. While few of us have flown a passenger aircraft or operated a nuclear power or chemical process plant, most of us have been patients – many in acute care hospitals. Healthcare issues affect all of us and our close families. Patient safety concerns touch everyone. And the basic issues present in complex socio-technical systems recur in healthcare organizations, often with some nasty variants. This domain has been my primary interest for the past dozen years or so.

The problem of patient safety is huge and it exists virtually everywhere. Early studies showed that more than 100,000 Americans die each year and that three million are injured as the result of medical errors.[1] Just over one in 10 British hospital patients are harmed by iatrogenic factors,[2] and roughly the same number, 10 per cent plus or minus two per cent, is echoed by the results of epidemiological studies in New Zealand, Australia, Denmark and Canada. A recent French study put the adverse event rate as high as 14.5 per cent of hospital admissions.[3]

Among the many reactions to these alarming numbers has been to look to other domains such as commercial aviation for lessons in safety management. While this is a very sensible measure, it must be appreciated that healthcare has many features that set it apart from aviation. Contemporary pilots – like nuclear power

1   Brennan, T.A., Leape, L.L. and Naird, N. (1991). 'Incidence of Adverse Events and Negligence in Hospitalized Patients'. *New England Journal of Medicine*, 324, pp.370–76.
2   Vincent, C. (2006). *Patient Safety*. Edinburgh: Elsevier.
3   Michel, P. et al. (2004). 'Comparison of Three Methods for Estimating Rates of Adverse Events and Preventable Adverse Events in Acute Care Hospitals'. *British Medical Journal*, 328, pp. 199–203.

plant and chemical process operators – are the supervisory controllers of largely automated activities. In aviation, their equipment is highly standardized and is mostly produced by the two principal manufacturers: Boeing and Airbus (each of whom has approximately half the world market in long-distance aircraft). The hazards of flight are reasonably well understood and the products of commercial aviation are delivered in a few-to-many fashion: two pilots and a small number of cabin crew can serve hundreds of passengers.

All of this stands in sharp contrast to healthcare. Here, both the activities and the equipment are exceedingly diverse, and while the latter may be less sophisticated than in aviation, its interpersonal dynamics are far more complicated, both psychologically and organizationally. Moreover, healthcare has more in common with aircraft maintenance than with the usually stable routines experienced by commercial jet pilots. Treating patients is a very 'hands-on' activity and, as such, is rich in error opportunities. In addition, the practice of medicine still has many unknowns and uncertainties. All of these features make both commission and omission errors more likely. And because patients are already vulnerable people, the likelihood of an error causing harm is much greater. Also, the hitherto localised investigation of adverse events makes it harder to learn and disseminate the wider lessons – unlike the detailed and publicly reported investigations of aircraft accidents.

Two other characteristics in particular set healthcare apart from other hazardous domains. In dangerous industries where the hazards are known and the operations are relatively stable and predictable, it is possible to deploy an extensive range of automated safeguards – or 'defences-in-depth'. Whereas some healthcare professionals (e.g., anaesthetists, intensivists and radiologists) use comparable automated safeguards, physicians, surgeons and nurses depend heavily on their own skills to keep patients out of harm's way. Patient injury is often just a few millimetres away. This brings us to the final distinction: the delivery of healthcare is a one-to-one or few-to-one business. Treating patients is a very close and personal interaction where the healthcarer's individual qualities are of supreme importance.

The chapter deals with the interactions between two dichotomies. The first is very familiar to readers – the person and system models of safety. The second is somewhat less familiar, being the oft-neglected distinction within the person model: the human-as-hazard and the human-as-hero. I have written at length about the latter elsewhere.[4]

Because the empirical foundations of the person model come mainly from event-dependent observations, it is inevitable that errors and violations are seen as dominating the risks to patient safety. But there is another side to the human factor and one that is often under-emphasized. Healthcare institutions would not function at all without the recoveries, adjustments, adaptations, compensations and improvizations made on a daily basis by healthcare professionals. As elsewhere, imperfect systems are occasionally pulled back from the brink of disaster by people on the spot. Eleven of these heroic recoveries are discussed at length in an earlier Ashgate book.[5]

In their more extreme forms, the person and system models of patient safety present starkly contrasting views on the origins, nature and management of unsafe human acts. They have been discussed at length in earlier chapters. A brief reminder: the person model sees errors and violations as arising from wayward mental processes, and focuses its remedial activities upon the erring individual. This view is legally and managerially convenient because it uncouples the responsibility of the individual from the organization at large.

The system model, on the other hand, views the front-line fallible person as the inheritor rather than the instigator of an adverse event. People at the 'sharp end' are seen as the victims of systemic error-provoking factors and flawed defences that combine, often unforeseeably, to bring about a bad event. The questions that follow from this model are not who screwed up, but which barriers and safeguards failed and how they could be improved to prevent a recurrence? I hope that it must now be clear that this is the view that prevails in this book.

---

4   Reason, J. (2008). *The Human Contribution*. Aldershot: Ashgate.
5   Reason, J (2008). *The Human Contribution*. Aldershot: Ashgate.

A Cyclical Progress

In what follows, we trace a patient safety journey that not only takes account of past and present developments, but also anticipates their future consequences. It begins during the 1990s with the widespread acceptance of the human-as-hazard aspect of the person model. It then takes us to the present, where a strong endorsement of the system model by many government reports has, among other influences, led to an increased awareness of error-causing factors acting at many levels of healthcare institutions.

However, it is also appreciated that systems are slow to change and we need to provide front-line staff with error wisdom – the mental skills that will help them identify and avoid high-risk situations. It is predicted that the adoption of these error management tools at the 'sharp end' will bring the human-as-hero to greater prominence.

But this can also carry a penalty. Local fixes may lead to the concealment of system problems from managers and others with the power and duty to effect more lasting global improvements. It is anticipated that when this process is better understood, there could be a backlash from managers, patients and lawyers that would bring about a reinstatement of the human-as-hazard view, albeit (we hope) in a more moderate form. At this point, the wheel will have come full circle.

Although these cycles will continue, it is hoped that healthcare institutions will learn and mature so that the wide variability evident in the initial go-arounds will gradually diminish to a state where all of these elements can co-exist harmoniously, leading to enhanced resilience and robustness. The main waypoints on this circular path are shown opposite in Figure 9.1. The letters A–D identify the temporal quadrants. Each quadrant is discussed separately below.

Quadrant A: From Human-as-Hazard to Awareness of Systemic Problems

This quadrant covers the period between the late 1990s and the present day. During these few years, low-level concerns about patient safety have escalated into a widespread acceptance that the problem is both huge and universal. In other hazardous domains,

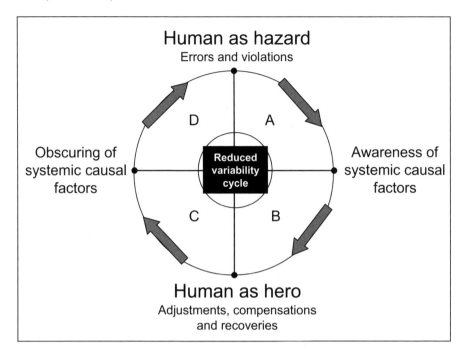

Figure 9.1   The cyclical progression of stages

such a dramatic change usually follows a well-publicized disaster, but, aside from some sentinel events, healthcare has had no 'big bangs' of this kind. Instead, the wake-up calls came from a flurry of influential reports and epidemiological studies. Perhaps the most influential of these was the Institute of Medicine's (IOM) publication, *To Err is Human: Building a Safer Health System,* which was released in the latter part of 1999.[6]

With some exceptions (particularly anaesthetists), the human-as-hazard view prevailed among healthcare professionals. This naming, blaming and shaming reaction to error was not only intuitively appealing, it was also strongly reinforced by a medical culture that equated fallibility with incompetence: the arduous, intensive and lengthy training was undertaken with the expectation of getting it right.

6   Institute of Medicine (2000). *To Err is Human: Building a Safer Health System.* Washington, DC: National Academy Press.

Quadrant B: Restoring the Balance between the System and the Person Models

Clearly, a systemic approach to the patient safety problem is a vast improvement over the simplistic human-as-hazard approach. But it is not without its drawbacks, as listed below:

- Those at the 'sharp end' of healthcare, particularly junior doctors and nurses, have little opportunity to carry out global system reforms. The following question therefore arises: how can we help them to cope more effectively with the risks they face on a daily basis? Could we not provide them with some basic mental skills that would help them to recognize and, if necessary, to step back from situations with high error potential (if it is possible)? In short, how can we enhance their 'error wisdom'? This question is discussed at length in Chapter 7.
- Such skills can also be useful for more senior clinicians and surgeons. These issues have been discussed in Chapter 4.
- Nurses, in particular, obtain a great deal of professional satisfaction from fixing system problems at a local level. However, as we shall see below, these workarounds can carry a penalty.

Quadrant C: The Downside of Human-as-Hero

At this point in the cycle, we move into the near future and the arguments become more speculative. Nonetheless, there are pointers available from current research to indicate how organizations might become disenchanted with too much individual autonomy, even of the heroic kind. One study in particular is especially compelling.

Tucker and Edmondson[7] observed the work of 26 nurses at nine hospitals in the US. Their primary interest was in the way in which the nurses dealt with local problems that impeded patient care.

---

7   Tucker, A.L. and Edmondson, A.C. (2003). 'Why Hospitals Don't Learn from Failures: Organizational and Psychological Dynamics that Inhibit System Change'. *California Management Review*, 45(2), pp. 55–72.

The problems included missing or broken equipment, missing or incomplete supplied, missing or incorrect information, waiting for a human or equipment resource to appear and multiple demands on their time. In 93 per cent of observed occasions, the solutions were short-term local fixes that enabled them to continue caring for their patients, but which did not tackle the underlying systemic shortcomings. Another strategy used on 42 per cent of occasions was to seek assistance from another nurse rather than from some more senior person who could do something about the root problem. In both cases, an opportunity for improving the system was lost. In addition, the nurses experienced an increasing sense of frustration and burnout, despite the often considerable personal satisfaction obtained from appearing to cope.

In Chapter 5, when discussing the fatal injection of vincristine (given spinally rather that intravenously), I introduced the phrase: 'the lethal convergence of benevolence'. Healthcarers do care very much about their patients and will sometimes bypass barriers or defences required by the protocol. In this instance, the protocol required that intravenous and intrathecal drug administrations should be given on separate days. But the patients, often teenagers, were not necessarily very good at keeping appointments. And the boy in this case was a notoriously bad attender. To overcome these problems, both the pharmacists and the nurses on the day ward appeared to have colluded to bypass the separate-days protocol. It was bad enough to get the patients to attend at all, but to attempt to get them to come twice was too much.

At a local level, these well-intentioned workarounds appear to smooth out many of the wrinkles of the working day. But from a wider perspective, it can be seen that they could carry serious penalties: the concealment of systemic problems from those whose job it is to remedy them and the bypassing or breaching of system safeguards. By their nature, these adverse consequences are not immediately obvious. In the short term, things appear to be working normally. This attitude of 'muddling through' is commonplace in complex systems. Yet, over time, latent pathogens are obscured and others are seeded into the system. This is an insidious process and it is only after a bad event that we appreciate how these separately seemingly innocent factors can combine to cause patient harm.

In addition, organizations that rely on – even encourage – these local fixes come to possess three inter-related systemic pathologies that are symptomatic of poor safety health:

- *Normalization of deviance*: this is an organizational process whereby certain problems or defects become so commonplace and so apparently inconsequential that their risk significance is gradually downgraded until it is accepted as being a normal part of everyday work. Such a process within NASA was cited as being a factor in both the Challenger and Columbia shuttle disasters.
- *Doing too much with too little*: this was another factor identified by the Columbia Accident Investigation Report as contributing to the Columbia shuttle tragedy. It is also a natural consequence of expecting front-line healthcarers to fix local problems as well as giving their patients adequate care.
- *Forgetting to be afraid*: because bad events do not appear to happen very often – at least from the limited perspective of individual carers – nurses and doctors can lose sight of the ways in which apparently minor defects can combine unexpectedly to cause major tragedies. If there is one defining characteristic of high-reliability organizations, it is chronic unease or the continual expectation that things can and will go wrong.

Quadrant D: The Reinstatement of the Human-as-Hazard Model

This is likely to occur at some time in the more distant future, so it is the most speculative of our transitional periods. Many of the processes described in Quadrant C are likely to be invisible to senior management on a day-to-day basis. It will probably take a number of well-publicised events implicating workarounds to bring them to light. But once their significance has been appreciated, it is almost certain that strong counter-measures will be introduced aimed at limiting the freedom of action of front-line carers. This backlash is likely to involve a number of top-down measures, the net result of which will be a return to the human-as-hazard model – at least in the eyes of management, though it will almost certainly be in a more moderate form. These measures could include the following:

- Bar coding, computerized physician order systems, electronic health records and automated dispensing systems have all been implemented to some degree over the past decade. Automation takes fallible human beings out of the control loop, at least in places where errors were commonly made. But this does not necessarily eliminate human error; it merely relocates it to the programmers and designers upstream of the front-line. It is probable that part of the backlash against human initiatives at the sharp end will take the form of more urgent attempts to computerize and automate human activities. In the past – and indeed the present – these innovations have been beset by technical and financial problems, but at this future time, it is possible that many of these difficulties will have been overcome. The history of complex hazardous systems tells us that one of the most popular methods that management commonly employ when dealing with human factors issues is to buy in what they see as hi-tech solutions.
- Another favoured counter-measure when dealing with the human factor is to write new procedures, protocols and guidelines that seek to limit sharp-end action to behaviours that are perceived as safe and productive. A fairly safe prediction, therefore, is that there will be intensified efforts to limit clinical autonomy. Procedures and guidelines have an important part to play in safety management, but they are not without their problems.

At first sight, it looks as though the cycle shown in Figure 9.1 has a 'good' sector (right-hand side) and a 'bad' sector (left-hand side). But each is a complex mixture: there is good in the 'bad' and bad in the 'good'. Nothing is wholly black or white: all have potential downsides and potential benefits. A better understanding of these issues will permit the anticipation and manipulation of their effects so as to maximize the positives and minimize the negatives.

Reduced Variability

As mentioned earlier, it is expected that as healthcare organizations learn more about these processes, variability over the cycle will diminish. The tensions and transitions implicit in the cycle will remain, but their perturbations will become less disruptive. It is hoped that the person and system models will operate cooperatively rather than competitively. This reduction in variability is represented in Figure 9.2

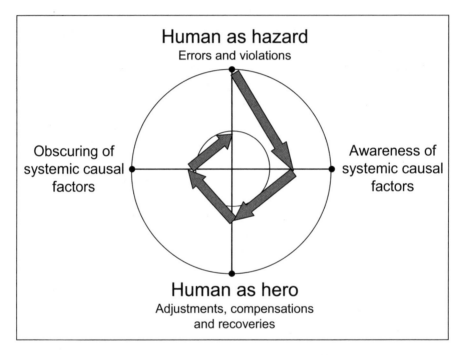

Figure 9.2   Reduction in variability on each successive cycle

It is not possible to step into the same river twice. By the same token, no organization remains the same. The inner circle in Figure 9.2 represents more moderate perspectives on the issues shown at the outer extremes. It is in this area of reduced variability that we would hope to achieve a mature balance between the system and person models and, within the latter, between the human-as-hazard and the human-as-hero distinctions. It is hoped that one of the enduring products of this equilibrium will be enhanced system resilience. We cannot expect to eliminate human error, technical problems and organizational pathogens altogether, but we can hope to create systems that are more resistant to their adverse effects. Greater resilience (unlike zero defects) is an achievable goal.

Mindfulness and Resilience

Increased automation of error-prone activities will do much to reduce many of the now commonplace error affordances in patient care, as will establishing the essential components of a safer organizational culture. But while they are necessary, they are not sufficient in themselves. Ultimately institutional resilience is dependent upon the mental and technical skills of those at the sharp end. Richard Cook and his co-authors argued that safe work in the real world depends critically upon:

> recognising that hazards are approaching; detecting and managing incipient failure; and, when failure cannot be avoided, working to recover from it.[8]

If there is one word that captures the essence of these skills, it is 'mindfulness', a term coined by Karl Weick and his colleagues at the University of Michigan. Mindfulness is needed at both the individual and organizational levels. Their relationship is represented in Figure 7.3 in Chapter 7.

The top half of Figure 7.3 shows an organogram for a healthcare institution. Front-line professionals are located on the interface between the organization and the patients. This is a very dynamic and turbulent interface, but in this picture, each carer is equipped with 'harm absorbers' or the mental skills necessary to cope with risky or error-prone situations. As the term suggests, 'harm absorbers' act like a vehicle shock absorbers – they iron out the perturbations on a rough road. System resilience is a function of this individual mindfulness of the dangers. But in order to be

---

8   Cook, R.I. and Nemeth, C. (2006). 'Taking Things in One's Stride: Cognitive Features of Two Resilient Performances'. In *Resilience Engineering*, edited by E. Hollnagel. Aldershot: Ashgate, p. 206.

effective, the harm absorbers need to be created and sustained by collective mindfulness – a property of the organizational culture that provides and trains the necessary mental skills and attitudes, and which empowers individuals to step back from what they perceive as dangerous situations and to seek more experienced assistance.

Organizational robustness, the institution's resistance to 'bumps in the road', hinges crucially upon the balance between the individual and collective mindfulness shown in Figure 7.3. In other words, it depends on a state of equilibrium between the person and system models of safety – a state that is iterated towards by the circular journey shown in Figures 9.1 and 9.2. I believe that this robustness exists when the organization has assimilated all the shades of opinion encountered on the route. Such an idealized state may still be a long way off, but this chapter has attempted to sketch out some of the waypoints and processes that are likely to be encountered en route.

Chapter 10
Taking Stock

Is Optimism Appropriate?

I ended the previous chapter on a somewhat upbeat note, but is any kind of optimism appropriate in the matter of organizational accidents (orgax)? To answer this, it is necessary to divide the issue into a hierarchical series of more focused questions.

To start at the top, will we ever be able to eliminate orgax? The answer, I fear, is an emphatic no. So long as we go on building and operating complex, tightly coupled hazardous systems, opportunities for unforeseen and often unforeseeable concatenations of failures that can defeat the defences will also continue to exist. And if there is a current trend, it is towards the development of more complex endeavours with increasingly remote centralized control. This will be spurred on by technological advances, particularly in automation.

So, if we can't prevent their occurrence, can we reduce their likelihood? Perhaps, but – and there is always a but – orgax, mercifully, are rare occurrences: they occur so infrequently and at such unpredictable intervals that if we did diminish them, we would probably never notice it. Maybe after a decade or two, we might wake up to their rarity.

Can we therefore improve a system's resistance to orgax? Here, I am inclined to offer a guardedly optimistic yes. My optimism is cautious because, in most domains, production invariably trumps protection. This precarious balance depends upon how long it has been since the domain suffered a high-profile disaster. But organizational accidents don't always provide the expected wake-up call. It is very unusual for an organization to suffer three organizational accidents in quick succession.

British Petroleum (BP) was the exception to this and I will outline these cases below. Before doing that, however, let me say that I visited BP facilities in Scotland and Victoria (Australia) in the early 1990s, and I was quite impressed with their safety sophistication. They even had their own home-grown version of Tripod. But the sad events related below tell another story. BP was riddled with unwarranted insouciance and hubris caught up with them.

- *The Texas City Refinery explosion:*[1] on 25 March 2005, a hydrocarbon vapour cloud explosion killed 15 workers and injured 170 others. BP was charged with criminal violations of federal environmental laws. The Occupational Safety and Health Administration gave BP a record fine for hundreds of safety violations and in 2009 imposed an even larger fine after claiming that it had failed to implement improvements following the disaster.
- *The Prudhoe Bay oil spill in Alaska:*[2] corroded pipeline owned by BP Exploration leaked 267,000 US gallons of oil over five days. It spilled over 1.9 acres, making it Alaska's largest oil spill to date. Alaska was fined $255 million; BP's share of the fine was $66 million. Warnings about corrosion had been raised several times both within and outside of the company.
- *Deepwater Horizon:*[3] a massive oil spill was caused in the Gulf of Mexico in 20 April 2010. It was due to a blowout on the BP-operated Manaconda platform. Eleven people died and it is rated as the largest accidental marine oil spill in the history of the petroleum industry. Oil flowed for 87 days and more than 200 million gallons of crude oil was pumped into the Gulf; 16,000 miles of coastline were affected. BP was fined close to $40 billion in clean-up costs and settlements, with an additional £16 billion due to the Clean Water Act. Countless wild animals were destroyed.

Did BP get its comeuppance on the markets? Not really. Its share price reached a peak in 2006 at around 712 p. It did, however,

---

1   BP Fatal Accident Investigation Report (2005) ibid.
2   See https://en.wikipedia.org/wiki/Prudhoe_Bay_oil_spill.
3   See https://en.wikipedia.org/wiki/Deepwater_Horizon_oil_spill.

have to sell off a large volume of shares to fund claims from the Gulf spill, though the total shareholder equity is still similar to its 2005 figure. In 2012, its net profit of $11.5 billion was half the $22.6 billion reported during 2005 and share price was down by around half too. Nevertheless, it currently has the ability to generate earnings similar to its 2005 levels. BP is one of the biggest integrated oil and gas companies in the world. To quote my dear and now sadly departed friend the Big Swede (Professor Berndt Brehmer), 'if you're big, you don't have to be nice'. That is one interpretation, at least.

Going back to the question posed at the outset, has the idea of orgax made a difference? So far we have concentrated on direct consequences – now let us consider some more indirect effects.

I was very gratified recently when another dear friend, Professor Najmedin Meshkati from the University of Southern California, sent me a copy of an opening statement made by the Acting Chairman, Christopher A. Hart, of the US National Transportation Safety Board (the premier accident investigation agency in the world). He was introducing a discussion of the tragic crash of Asiana Airlines Boeing 777 at San Francisco Airport on 6 July 2013. Three people died and 87 people were injured (47 seriously), due, among other things, to a misuse of automation. The aircraft struck a seawall on approach, the tail broke off and an engine caught fire. Here is the quote:

> More than 15 years ago, Professor James Reason wrote: 'In their efforts to compensate for the unreliability of human performance, the designers of automated control systems have unwittingly created opportunities for new error types that can be even more serious than those they were seeking to avoid'. Many others had made similar observations around that time.[4]

Assisting in the interpretation of accident causes is a rather remote effect, though not necessarily trivial, but it did stem directly from the idea of orgax. And it is, of course, a very un-English blast on my own trumpet. But how could I resist its mention? There are so few other clear acknowledgements – except, perhaps, in healthcare. Nonetheless, sales of the book indicate that many

---

4   Najmedin Meshkati (personal communication).

people may have read it – it was for a while Ashgate's best seller
(at least on the human performance list).

Full Circle?

In every hazardous system, there is some degree of conflict
between production (that which contributes to the bottom line)
and protection (the activities that seek to keep the system safe).
There are many reasons for this. First, production pays for
protection, except when it fails and an accident causes a serious
dent in the bottom line. Second, the people who manage the system
are usually more trained in productive measures than protective
ones: they are the ones striving to meet the profit targets; success
or otherwise in this regard will have an immediate impact on
their careers. And, third, the effective management of protection
– safety management – is extremely difficult. No one can foresee
all the ways in which harm can enter a system. The immediate
dangers can be very apparent, but the various ways in which
resident pathogens can combine with human activities and the
local conditions are often unexpected and unforeseen. Resident
pathogens exist in all systems; they are universal conditions.

So if all hazardous systems have production and protection
conflicts, and if all of them suffer resident pathogens, what is
it that distinguishes – within a particular domain – the system
having an organizational accident from those that do not? All
face the same long-term problems to some degree or another;
as such, the only things making a causal difference can be the
local circumstances prevailing on the day (or the immediately
preceding days) of the accident. Could it be that I have come full
circle to blaming the unsafe acts of those at the sharp end? No, that
would be a gross error – understandable, but gross nonetheless.
Why?

Consider the capsizing of the *Herald of Free Enterprise* ferry.
Several factors were at work on the day of sailing, but perhaps
the most important was the month: March. Due to the high spring
tide and equipment problems at Zeebrugge, the ramp could not
be elevated sufficiently to permit car loading, so the ship was
trimmed nose down and continued to be so when it left harbour.

Errors are consequences rather causes. To illustrate this, it is worth considering what Mr Justice Sheen, the Wreck Commissioner, wrote in his report on the accident:

> At first sight the faults which led to this disaster were the errors of omission on the part of the Master, the Chief officer and the assistant bosun ... But a full investigation ... leads inexorably to the conclusion that the underlying or cardinal faults lay higher up in the Company ... From top to bottom the body corporate was infected with the disease of sloppiness.[5]

The *Herald* was a sick ship before it even reached Zeebrugge. It lacked bow-door indicators, showing that doors were open or shut – despite repeated requests from the crew for this. It was undermanned and had a chronic list to port. It took relatively little in the local circumstances to bring about the disastrous capsizing.

Local factors are what distinguish the system that suffers an accident from those that do not. But these local factors, especially unsafe acts, often have their origins in upstream resident pathogens. The Master, being unaware that the bows doors were open, left the harbour at his usual speed, which, due to the nose-down trim, was sufficient for seawater to enter the boat deck and cause the capsizing. Had the ship been on an even keel, there would have been a three-metre gap between the sea and the open doors; with the nose down trim, however, this was reduced to one metre.

Counter-factual Fallacy

The Columbia Accident Investigation Board's report into the Columbia shuttle disaster contained at least two examples of this fallacy:

> In our view the NASA organizational culture had as much to do with this accident as the foam. [It will be recalled that a large piece of foam penetrated the leading edge of the Shuttle's wing.]

5   Mr Justice Sheen (1987). *MV Herald of Free Enterprise*. Report of Court No. 8074. London: Department of Transport, p. 14.

> The causal roots of this accident can be traced, in part, to the turbulent post-Cold War policy environment in which NASA functioned during most of the years between the destruction of Challenger and the loss of Columbia.

During these years, NASA was directed by a cost-cutting, downsizing, 'leaning and meaning' administration. But these were the Thatcher and Reagan years; they were the buzzwords of the age. Everyone was exposed to them. They are conditions, not causes.

The counter-factual fallacy goes as follows. All accident investigations reveal systemic shortcomings. It is but a short step, therefore, to argue that these latent conditions caused the accident. Yet this is not necessarily correct. There are always organizational interventions that could have thwarted the accident sequence, but their absence does not demonstrate a causal connection. So the fallacy is this: if things had been different, then the accident would not have happened; ergo, the absence of such differences caused the accident. However, these organizational factors are conditions, not causes – necessary perhaps, but insufficient to bring about the disaster. Only the local circumstances are necessary *and* sufficient.

Systems Thinking

Throughout this book, I have been airily tossing around the term 'system', usually as a term interchangeable with organization. But it is nowhere near as simple as that. A system is something larger than any one group of workers or one particular site or department. It has its own internal processes and logic. In order to redress the balance, I am going to turn to the system theorists, of whom there are many; however, the one I have found most helpful is Jake Chapman of the Open University, who has advised governments and corporations as well as teaching systems theory. I am especially indebted to his excellent book (or pamphlet as he modestly calls it) *System Failure*.[6] I have to warn you that systems theory runs counter to many cherished beliefs, the most important of which is mechanistic and reductionist thinking. For

---

6   Chapman, J. (2004). *System Failure*, 2nd edn. Milton Keynes: Demos.

most of us who have received a technical or scientific education, the accepted way of solving problems is to break them down into discrete elements (reductionism) and to find solutions to these bits one by one (this is what we call the 'evidence-based' approach). This stance is deeply embedded in our culture. It assumes a linear relationship between cause and effect (Hollnagel has taken us here before).

Our primary concern is with the complex hazardous systems that have occupied us in this book, but these complex systems involve a multiplicity of nested feedback loops which can lead to significantly nonlinear behaviour. This means that cause-and-effect processes, far from being linear, can loop back on themselves. One of the more predictable outcomes of inappropriately applying linear thinking to complex systems will be unintended consequences, of which there are many instances in healthcare. Here is one example:

> The Nation Audit Office (NAO) found that in order to avoid being fined for over-long waiting lists, 20 per cent of consultants re-ordered priorities in their operation lists favouring simple routine procedures over more complex ones in order to achieve their numerical targets.

It has been pointed out that, especially in complex adaptive systems, the pursuit of single quantified targets is counter-productive. It lowers quality, raises costs and distorts systems, reducing their overall effectiveness. Forcing systems to prioritize one aspect of their performance will distort their general performance and will thus impoverish the broader aspects of their activities. Chapman adduces many examples to suggest that the only effective judge of performance is the end-user.

One example of overcoming imposed measures of performance is the Toyota production system (TPS). The TPS has been refined and developed over the past 50 years using the same technology and general principles as other car manufacturers. 'Greater capability' within the TPS is not assessed by any single measure; it is an aggregate of many measures. The key to the success of the TPS is its attention to detail, its active involvement of those at the sharp end, and the belief that sustained improvement can only be achieved over a long period by incremental progress.

This approach conflicts with the requirements of politicians and senior managers who make promises to improve schools, reduce waiting lists and to cut accident rates by x per cent over the next N years. Even worse, I believe, is the promise or desire to eliminate accidents altogether. This 'target zero' attitude only misrepresents the nature of the safety war – which is a guerrilla conflict where there will be no clear victories like Waterloo or Appomattox.

While I believe that systems thinking can play an important part in the management of organizational accidents, I have my concerns. The first is that few of the managers of hazardous systems will be graduates of the systems courses at the Open University or other comparable universities. The second is that while the dangers of mechanist-reductionist thinking are clear, they also have their uses. I'm deeply suspicious of polar views (i.e., systems versus linear approaches) and am a firm believer in the notion that there is no 'one best way' in our complex world. But what worries me most about system thinking is that while problem diagnosis may (sometimes) be fairly straightforward, effective treatment seems uncertain and unacceptably vague.

Chapter 11
Heroic Recoveries

The Miracle on the Hudson (2009)

No account of heroic recovery is complete without mention of Captain Chesley B. Sullenberger III ('Sully') and his First Officer, Jeffrey Skiles. They were flying an Airbus A320-200 out of LaGuardia Airport in New York City on 15 January 2009. About three minutes into the flight at 3.27 pm EST, the aircraft struck a flock of Canada geese during its initial climb out. The bird strike caused both jet engines to quickly lose power. As the aircraft lost altitude, the pilots decided they could not return to LaGuardia or reach Teterboro airfield in New Jersey. They turned southward and glided over the Hudson, finally ditching the airliner off midtown Manhattan into the Hudson River about three minutes after losing power.

There was little verbal communication between the pilots. First Officer Skiles was performing the take-off, but immediately after the bird strike Captain Sullenberger took over the controls. 'I have the aircraft' he said. Very shortly after, as they were lining up for the ditching, he asked the First Officer to put down the flaps.

The ditching was a miracle of airmanship. The aircraft was descending in a slightly nose-up attitude. Had the nose been lower, it could have ploughed into the river and ended up on the bottom. Had the tail been lower, the same fate would have resulted. As the plane hit the river and skimmed along the surface, water streamed over the cockpit windows. It finally settled, more or less horizontal, whereupon it drifted downstream with the current. All 155 occupants, the passengers and crew, successfully evacuated from the partially submerged airframe as it sank into the river. They were picked by rescuers that included ferries, tugboats, Coast Guard vessels and others who through their prompt arrival and braving the deadly cold water were able to fish out the passengers and crew.

Before leaving the aircraft, the Captain walked up and down the passenger area, twice checking the passenger seats. The aircraft was filling with water. At the NTSB hearings, Captain Sullenberger talked of a 'dedicated, well-experienced, highly trained crew that can overcome substantial odds and work together as a team'.

Fukushima Daini (2011)

I will conclude with a description of Fukushima Daini – sister plant to the one that had the disastrous meltdowns in 2011. The Daini story[1] is one of heroic recovery and brilliant leadership. My other reason for including it is to show that mechanistic-reductionist thinking can have its uses.

The Fukushima Daini plant was 10 km to the south of Fukushima Daichi and was exposed to the same earthquakes and tsunamis, but with no meltdowns (unlike Daichi – as the world knows). During the four-day period of the emergency, it was exposed to around a dozen perturbations, nearly all of them requiring reappraisals of the recovery strategy. Once again, I am eternally grateful to Najmedin Meshkati for sending me a *Harvard Review* article.[2]

On 11 March 2011, an earthquake with a magnitude of 9.0 struck the Japanese coastline. It was the largest in Japan's recorded history. The associated tsunami generated waves that were three times as high as that which Daini had been built to withstand. A timeline of events is shown below:

11 March
- 2.46 pm: a huge earthquake and tsunamis strike. This is followed by a succession of aftershocks. Workers leave the

---

1   National Research Council of the National Academies (2014). *Lessons Learned from the Fukushima Nuclear Accident for Improving Safety of US Nuclear Plants*. Washington, DC: National Academies Press.
2   Meshkati, N. (2014). Personal communication.

administration building and squeeze into the Emergency Response Centre.

- 3.22 pm: the surge of water knocked out the power in the Emergency Response Centre. The plant is left with a single power line to the control rooms and one emergency diesel generator at Unit 3.
- 6.33 pm: three of the four reactors lose their cooling abilities. Fuel rods inside each core continue to generate heat.
- 10.00 pm: small teams of workers go outside to assess the damage in the field. Most live near the plant and are desperate for news of their families.

12 March
- Early morning: Unit 2 is judged as being at greatest risk. The site supervisor, Naohiro Masuda, maps a route for cables to draw power from the radioactive waste building.
- During the day: workers manhandle huge cable sections so that ruined pump motors can be replaced and damaged parts of the cooling system can be connected to the radioactive waste building – the one still drawing power. They have to lay more than 9 km of cable to link up the three disabled units. They have about 24 hours to complete a job that normally took a month. Later Masuda recognizes that the plan to use the waste building as a source of power was impractical due to its remoteness, so he reluctantly uses the single diesel generator.

13 March
- Throughout the emergency, Masuda had been sharing his ideas with his workers using a whiteboard, slowly replacing uncertainty with meaning: what Karl Weick has called 'sense-making'. This was a major factor in the success of the recovery. Not only did he have to keep the workforce up to speed with each new problem as it arose, he had to persuade them to act against their survival instincts. A broken nuclear plant is a very dangerous place.
- Midnight: Masuda learns that another reactor, Unit 3, was at the greatest risk of meltdown. The workers finish re-routing the cable.

14 March
- 1.24 am: cooling function is restored to Unit 1.
- 7.13 am: cooling is restored to Unit 2.
- 3.42 pm: cooling is restored to Unit 4.

15 March
- 7.15 am: all four reactors are in cool shutdown.

What Led to the Successful Recovery?

A crucial factor was the inspired leadership of Naohiro Masuda. Another was the stoicism and heroism of the workforce, particularly those who ventured outside. Together they charted their way through the crisis, taking nothing for granted and dealing with each new problem as it arose. Collectively they acknowledged the evolving reality in which they were operating and iterated their way to a better understanding.

Masuda had worked at the plant for 29 years; he knew every nook and cranny. This technical competence and his diligence and openness helped him earn the trust of the workforce. He arrived at a common understanding with his team members and communicating what they 'knew' so that together they could adapt to each twist and turn.

Fittingly, he was appointed Daichi's (the sister plant) Chief Decommissioning Officer in April 2014.

The Ingredients of Heroic Recovery: What is the 'Right Stuff'?

What makes for heroic recoveries? This is a question I have struggled with at length in a previous book. The short and unsatisfactory answer is that it involves the right person in the right place at the right time doing the right thing. I don't believe that this statement is untrue, but it is nevertheless unhelpful because it relates to past emergencies. How can we combine these ingredients to cope with a future event? The only way I know of doing this (against the dictates of system theory) is to break down these elements into their discrete parts. Perhaps the most tractable of these is the 'right person'. By what criteria do we select such a person? It seemed to me that the most obvious place to look first was at the criteria used by elite military units to select officer candidates for such forces as the US Marine Corps and the UK Special Services (for example, the SAS).

It didn't take me long to realize that I was looking in the wrong place. Naturally enough, they were seeking young, vigorous (if not athletic) boys and girls with leadership qualities – which one would assume had been established in their two to three years of arduous officer training. But none of these matched the template of my proven heroic rescuers – all of whom, bar one, were maturely middle-aged with a wealth of experience in their particular technologies and who, most importantly, were able to think intelligently on their feet. A great plus was that all of them were confident in their abilities and were able to communicate this to their associates. Above all, they were blessed with realistic optimism: the feeling that it would come out alright in the end. This too they were able to pass on. Despair is the greatest enemy of heroic recovery.

End Piece

I cannot end without once more expressing my enormous indebtedness to Professor Najmedin Meshkati and his co-author, Yalde Khashe. Their paper, 'Operators' Improvisation in Complex Technological Systems: Successfully Tackling Ambiguity, Enhancing Resiliency and the Last Resort to Averting Disaster', was published in the *Journal of Contingencies and Crisis Management* (2015). In 2008, I wrote a book entitled *The Human Contribution: Unsafe Acts, Accidents and Heroic Recoveries*. Their paper goes well beyond what I wrote there or had thought about. I receive emails from Najmedin on an almost daily basis. I hope he never stops and I thank him again for his generosity of spirit and wonderful scholarship.

Index

Note: Figures and tables are in a **bold** font.